PASSIVE SOLAR ARCHITECTURE

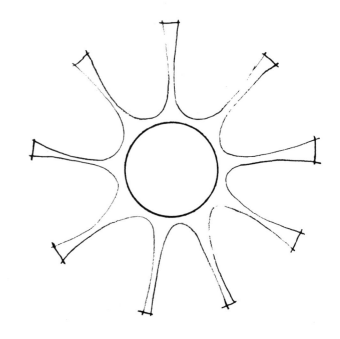

PASSIVE SOLAR ARCHITECTURE:
logic and beauty

35 Outstanding Houses Across the United States

David Wright, AIA
Dennis A. Andrejko, AIA

Phill Cooper   Illustrations
Steven R. Solinsky   Photographs
Raymond K. Darby, Jr.   Computer Analysis

VNR Van Nostrand Reinhold Company
New York Cincinnati Toronto London Melbourne

Printed in the United States of America
Designed by Loudan Enterprises

Published by Van Nostrand Reinhold Company Inc.
135 West 50th Street
New York, NY 10020

Van Nostrand Reinhold Limited
1410 Birchmount Road
Scarborough, Ontario M1P 2E7, Canada

Van Nostrand Reinhold Publishers
480 Latrobe Street
Melbourne, Victoria 3000, Australia

Van Nostrand Reinhold Company Limited
Molly Millars Lane
Wokingham, Berkshire, England

16 15 14 13 12 11 10 9 8 7 6 5 4 3 2 1

Library of Congress Cataloging in Publication Data

Wright, David, 1941–
    Passive solar architecture.

Bibliography: p. 252
    Includes index.
    1. Architecture and solar radiation—United
States.   2 Architecture, Domestic—United States.
I. Andrejko, Dennis A.   II. Title.
NA7117.S65W7   728.3′7′0470973   82–2862
ISBN 0-442-23860-6   AACR2
ISBN 0-442-23859-2 (pbk.)

ACKNOWLEDGMENTS

Our sincere thanks and appreciation to the following
for help with this book:
    For technical review and critique of the rough draft,
Skip Benoit, solar representative, Pacific Gas and
Electric Company; Jeffery Cook, professor, College of
Architecture, Arizona State University; Fred Nelson,
editor, *Sunset* magazine; John Yellott, professor
emeritus, College of Architecture, Arizona State
University.
    For the loving support of our families, Susan, Erik,
and Bryan Andrejko, Cathy and Megan Wright.
    For dedicated and talented word processing, Nancy
Wolters.
    For their patience and support, The SEAgroup.
    For the computer program description in Appendix
A, The Berkeley Solar Group, Ray Darby.
    For their words and architecture, all of the
contributors herein.

PHOTO CREDITS

All photographs other than those credited below were taken
by Steven R. Solinsky, Nevada City, California.

page 9 (top), anonymous; page 9 (bottom), David Wright;
pages 58–59 (top), page 62, Tim Street-Porter; page 64,
Lawrence Hudetz; page 65, William Church; pages 76–78,
James Lambeth; page 125, Janus Associates, Inc.; page 148,
Peter Pfister; pages 177–78, Robert Perron; pages 196–200,
Rick Rusing; pages 204–08, Steve Rosenthal; pages 239–43,
Kristen Peterson.

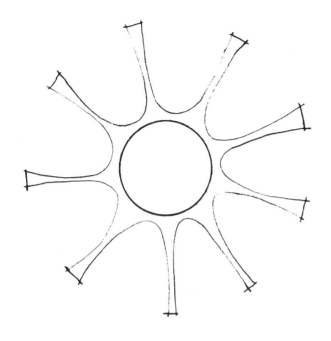

CONTENTS

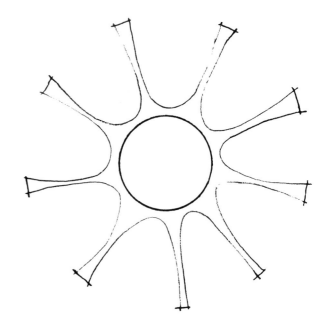

PREFACE

The energy crisis continues, and avid proponents of solving energy-consumption problems press on with new solutions. In architecture, new designs and systems are rapidly being developed, proposed, and built. Often the wheel is reinvented. Sometimes original solutions appear. Too often, complex answers are offered for simple needs, although the economics of energy use usually dictates simple solutions. As architects employing energy conservation as a primary aspect of design, we believe in applying sound design basics before searching for esoteric solutions. The common-sense rules of thoughtful architecture are primary in designing efficient buildings.

We have sensed a common denominator among designers facing the challenge of energy-conscious architecture. We have tried to identify the elements—cultural, geographical, and philosophical—that comprise the bold emerging architectural approach to today's needs. Our search across the United States revealed an undercurrent of exciting and talented people designing better buildings that use solar power in the most logical, cost-effective way. We found no single stylistic thread, but, instead, diverse architectural styles that share passive solar design techniques. The common denominators include solar orientation; thermal tightness, storage, and control; solar glazing; and innovative detailing to accommodate these architectural aspects.

Our final judgment is that a new wave of architectural awareness is sweeping the country. Future architecture of the United States, and indeed of the world, will bear the imprint of the logical and beautiful passive solar design features that are now in the adolescent stage of development. We are proud and delighted to be a part of this architectural frontier.

Chapter 1

LOGIC AND BEAUTY

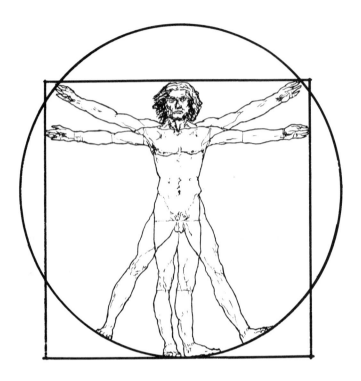

Successful architecture has always been a celebration of logic and beauty. The logic involves function, structure, economics, planning, and the myriad other aspects of the architectural arts. The beauty is truly in the eye of the beholder, whether owner, user, architect, builder, neighbor, community, or society. Inescapably, neither logic nor beauty is complete without the other.

Mankind has always aspired to create buildings that fulfill the yearning for these themes. The timeless examples of successful work always incorporate the environment's natural energy flows. Persian garden villas, Chinese walled cities, Japanese homes, Greek temples, Roman baths, American Anasazi Indian cliff dwellings, and most other architectural examples from developed cultures incorporated the earth, wind, and fire of nature's flux in their design. And the use of solar energy in all its natural forms gave a unique quality to each architectural style. A careful study of architectural history reveals that past cultures understood the potentials of on-site energy flows to heat, cool, ventilate, and light their buildings. The most outstanding of these buildings were truly delightful examples of man's ingenuity in solving problems and adapting natural phenomena to the art of building.

Unfortunately, man is basically lazy. Modern architecture has favored expediency and a fascination with technology. Western man in particular often lets machines handle a maximum amount of work to make life apparently easier. Easiest, however, is not always best in the long run. During the past half century, with the availability and intense use of fossil fuels, the building industry and its clients have taken maximum advantage of machines to control light and temperature and comfort in their buildings. Only since the 1973 oil embargo has it become apparent that this approach may not be the wisest. The energy crisis is real and continues. It is now

The natural site provided summer cooling and winter heating for Anasazi Indians, ancestors of the Pueblo Indians, in the rugged southwestern United States.

clear that some of the most highly praised architectural solutions of the last few decades, particularly for small structures and the family home, lack efficient space conditioning.

Much of the knowledge of the past was left forgotten on library shelves in the quest to adopt air conditioning as the throbbing heart of modern buildings. Too often, architects dealt only with the structure, form, and function of the design, usually specifying that the mechanical-system designer plug in a unit and not ruin the esthetic appeal with visual clutter, discomfort, or poor performance. Many contemporary architects have failed to remember, or never learned, the basic, common-sense aspects of natural controls. They have indulged themselves and saddled clients with buildings that cannot function independently of space-conditioning machinery. When power is interrupted, limited, or too costly, many modern buildings become uninhabitable. Overdependence on the wall thermostat has sometimes even elimi-

Contemporary pueblo-style architecture echoes the knowledge of the ancients.

nated effective fenestration and other links with the outside. In short, buildings have been produced that override, rather than complement, nature.

The energy crisis has caused architects and engineers to reconsider design priorities and to deal realistically with the economics of energy use. Exponentially accelerating energy costs, dependence on imported fuels, and a need to use energy efficiently have brought about the beginning of energy-conscious architecture. The designs of the future will, of necessity, be able to ensure optimum comfort with a minimum of energy dependence.

This rational new architecture is blessed with much more than fuel frugality. As designers endeavor to solve energy problems, they are rediscovering, as well as inventing, architectural elements that not only work better, but that create a new esthetic. Logic and beauty are the threads that interweave throughout the work that is beginning to manifest itself. Passive solar design—using the sun to heat and cool buildings by nonmechanical means—is one element. Passive solar architecture has proven to be highly efficient, cost-effective, and creatively stimulating.

Successful architectural design never takes an inflexible approach to energy efficiency or any other single design element. Rather, all design factors must be satisfied and mutually compatible. Thus, creating energy-conscious buildings involves an awareness of the traditional factors of common-sense design, as well as the new, practical concerns. By satisfactorily utilizing all the variables of the architectural palette, the designer can create a new esthetic statement. The factors generating the emergence of energy-efficient architecture will help to spawn a fresh design freedom for responsible architects.

This book documents the work of a few of the designers and architects who are addressing the issue of solar architecture. These examples are presented to illustrate the special logic and beauty of energy-conscious passive solar architecture. The new and retrofitted residential projects presented in Chapters 2 through 6 are the results of a new design ethic. This book contains basic information conveying the essence of the special features of each design shown. In certain cases, the inhabitant's statement or the computer analysis is not included because it was not available. These examples have been selected to illustrate some of the ideas, methodologies, and esthetics that are a result of aware individuals adapting new concerns and tools to the ancient art of building design. It is clear that this is only the beginning of a new era of resourceful design; future architects will deal with more difficult challenges. For the present, the new direction is eminently exciting.

Climatic Analysis

In order for a passive solar building to be successful, it is essential for it to interact with the climate. This requires an understanding of climatic factors at the following three levels:

1. Macroclimate, or large scale—regional climate conditions
2. Microclimate, or local scale—external environment at the site
3. Interior climate, or human scale—conditions within the building

Man's survival is governed by his ability to adapt at each of these three levels. The architect can design a building that integrates as well as possible with the environment, or create an artificial place in which to exist. Sometimes he does both. Often the solution lies somewhere between.

Macroclimate

The macroclimate consists of the general climate of the region in which a building operates. The macroclimate is determined in part by latitude, elevation, and general terrain. Regional design techniques and local customs should, and often do, take these patterns into consideration. New England winters, for example, are known for deep snows accompanied by chillingly low temperatures that can last many months. People there traditionally wear heavy boots and socks, jackets, and layers of warm clothing to protect their metabolic furnace. The architectural vernacular, or regional building style, reflects this approach: the typical New England–style homes, like the Cape Cod and the saltbox, have an inner core, often a masonry fireplace, around which spaces with the highest heat demand are nestled. An additional layer of spaces, with lower demand for heat, is located on the outside perimeter, acting as a buffer; this layer often comprises storage, utility, and bathroom areas. The ratio of the exterior building surface area, or weatherskin, to the interior building volume is kept low. This minimizes the heat loss surface of the living area.

On the other hand, the need for heating is very low or nonexistent in the southern bayou country along the Gulf of Mexico. This region has frequent and heavy rains, hot, humid, sticky summers, and fairly mild winters. People dress in loose cotton clothing with open neck and sleeves. The need for surfaces of both people and buildings to "breathe" and take advantage of air movement is particularly important in this climate. The regional architecture has therefore adopted high ceilings with floor-to-ceiling openings that allow cross-ventilation and screened shaded porches; umbrella roofs, high central galleries, operable vents, shade roofs, and ventilating cupolas or dormers are other architectural features that offer successful passive climate control for this region.

The accompanying map has been

United States Climate Map

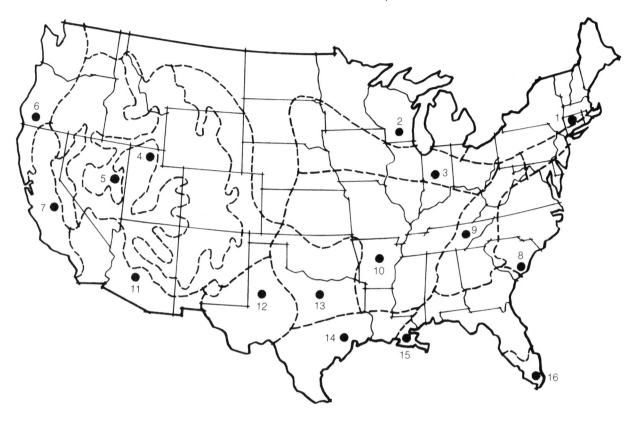

1. Hartford, CT
2. Madison, WI
3. Indianapolis, IN
4. Salt Lake City, UT
5. Ely, NV
6. Medford, OR
7. Fresno, CA
8. Charleston, SC
9. Knoxville, TN
10. Little Rock, AR
11. Phoenix, AZ
12. Midland, TX
13. Ft. Worth, TX
14. Houston, TX
15. New Orleans, LA
16. Miami, FL

1. *Northeast*—Harsh winters, warm humid summers
2. *North*—Extreme cold winter, warm summers
3. *Midwest*—Cold winters, hot humid summers
4. *Great Plains*—Cloudy windy winters, warm dry summers
5. *Mountainous west*—Very cold winters, mild dry summers, large day-night temperature swings
6. *Northwest Coast*—Consistently cool temperatures, rain and fog, extreme microclimate variation
7. *Southwest Coast*—Moderately cold rainy winters, hot dry summers, reliable sunshine
8. *Mid-Atlantic Coast*—Cool to cold winters, warm summers
9. *Appalachian*—Cool winters, hot humid summers
10. *Mississippi Valley Region*—cold winters, hot humid summers
11. *Southwest Desert*—Extremely hot summers, moderately cold winters
12. *West Texas Area*—Cool dry winters, hot dry summers, dependable sunshine
13. *North Texas Area*—Potentially cold winters, hot summers with variable humidity
14. *Gulf Coast*—Very mild winters, hot humid summers, substantial sunshine
15. *Bayou Coast*—Wet cool winters, excessively hot humid summers
16. *Tropical Southeast*—Very mild winters, balmy sunny summers

adapted from the HUD publication *Regional Design Guidelines for Designing Passive Energy Conserving Homes.* It identifies U.S. regions by climate type and provides an initial means of assessing climatic assets and liabilities, helping the designer to understand the impact of a particular macroclimate.

Architectural regionalism usually offers a design type that is naturally suited only to a particular climate and would not be suited to another without selective modification. An understanding of the local climate enables the architect to adapt a building to its immediate environment.

Microclimate

Each time a building is built, man modifies the landscape. At the design stage, the architect makes fundamental decisions that determine to what degree this intrusion will influence the environment. Whether the locale is urban, suburban, or rural, the designer should study several descriptive indicators of site and climate, as well as considering planning and building factors. These reveal to the designer the proper siting, orientation, form, materials, and other suitable aspects for each particular design solution. A detailed evaluation of the indicators, which may be very subtle, can be used to determine the impact of a particular design.

Site characteristics vary dramatically from place to place. A coastal site, for example, can vary from beach to cliff to meadow over a short distance, creating quite different characteristics, even though the macroclimate is uniform. Weather characteristics also vary but these are more difficult to assess. There are few locations where thorough local weather information is available, and there are practically no accurate means of interpreting regional climate data into precise microclimatic data. The more local the data, the better. Neighbors, farmers, chambers of commerce,

agricultural agents, and airport operators are all valuable sources of this information.

A thorough microclimatic analysis will suggest to the designer ideas for form and material. Without this analysis, there is no sound basis for making an environmentally attuned decision: the architect is relying only on the man-made or externally derived determinants of architectural system, style, or economics. Designing with the microclimate in mind puts the designer on the side of nature, so that man's influence is complementary to the elements that already exist.

The microclimate matrix illustrates some of the issues that the building designer should be adept in assessing and interpreting before commencing any energy-conscious design process.

Interior Climate

The last level of climate is that within the building itself. Since the ultimate goal of any house design is to provide shelter and human comfort, it is necessary to know how and why we experience comfort in different circumstances.

The human body is essentially a low-grade furnace that uses food and fluids as fuel. In order to maintain an average interior metabolic temperature of 98.6°F (37°C), this furnace must function well throughout a wide range of climates and physical activities. Depending upon climate and activity, the metabolic rate, or rate of energy lost from the body, varies. For example, an adult reading a book loses approximately 300 to 400 British thermal units (Btus) per hour. A sleeping adult loses 250 Btus per hour; one performing heavy work loses over 2,000 Btus per hour.

People (wearing normal clothing) are most effective at maintaining proper body heat loss when the external environment is about 25°F (−3.8°C) cooler than our body temperature. This is a temperature dif-

ferential at which the body operates quite efficiently. How the body loses heat specifically depends upon the thermal conditions of the environment, but it is generally the same way a building does. This includes:

Conduction—heat transfer from a warmer surface to an adjacent, cooler surface by physical contact. Applying an ice pack rapidly cools the skin.

Convection—heat transfer by or within a moving fluid (liquid or gas) from a warmer to a cooler surface. A breeze carries heat and moisture away from our body surface by convection.

Radiation—heat transfer by electromagnetic waves from a warmer to a cooler surface. The sun warms us by radiation, and we lose body heat to a cold window at night by the same effect.

Evaporation—the dissipation of heat when moisture on a surface is vaporized. Perspiration being given off to the surrounding air is an example of heat lost through evaporation.

Convection and radiation are the two primary means by which the body loses heat. At 60°F (15.5°C), the body loses about seven times more heat by these methods than by evaporation. Little heat is lost from the body by conduction, since usually only a small portion of the body is in direct contact with another surface. Transfer by evaporation, especially at air temperatures of 60–80°F (15.5–26.5°C), is slight. Only above this range and with low humidity does transfer by evaporation become significant.

The interior thermal environment of our living spaces can be regulated in several ways, physical and psychological, to allow our bodies to lose heat comfortably. The following factors should be considered to achieve proper environmental comfort:

air temperature
mean radiant temperature (MRT)
relative humidity (RH)
air motion, including speed, volume, direction, and type

Microclimate Matrix

Issue	Effect/Need	Evaluation Method
Site Characteristics		
Soil conditions	Subsurface drainage, percolation, bearing value, structural uses, underground stability, heat/coolth capacity.	Visual examination. Soil test sample. Back-hoe trenching.
Profile	Surface drainage, solar exposure, shadowing, weather patterns at the site, earth integration construction suitability, wind power.	Compass and site level, transit site survey, and contour map.
Vegetation	Planting or removal to enhance a microclimate; work with existing as pertinent; indigenous species for suitability.	Visual examination, samples, landscape architect, log existing trees, plant/shrub ground cover, and grasses.
Materials	Potential use of inexpensive, low energy, indigenous resources. Generally suited to the climatic elements.	Visual examination and samples.
Water	Well water, water recovery/greywater, evaporative cooling and heat sink potential, purification and treatment.	Utility maps, visual examination, test drilling, water witching, water test.
View	Orientation, fenestration.	Photography, visual examination.
Climate Characteristics		
Sunlight (or lack of it)	Most important factor. Determines air temperature, air movement, precipitation, vegetation characteristics, and most aspects of building design (orientation, fenestration, building profile, solar suitability, etc.).	Determine solar access with solar site selector device. Solar availability from climate data sources to include hours of sunshine, percent of possible, and type of direct or diffuse radiation available. On-site radiation measurement with pyranometer.
Air motion and wind	Encourage wind currents when needed, isolate from when a liability. Can be used effectively to reduce heating, cooling, and ventilation loads. May require earth integration or building profile and orientation modification to reduce infiltration. Wind shadow location for outside use areas. Storm direction.	Vegetation characteristics and land profiles will give indication of prevailing wind directions. Climate data for night, morning, and evening. Speeds and directions of wind, national weather records, local agricultural or aviation records, neighboring farmers, commercial records, etc.
Temperature	Needed to determine design temperatures, skin type, insulation quantity, comfort parameters, building operation patterns.	Recording thermometer for site analysis. Climate data for mean, maximum, minimum, day and monthly, also dew point. Heating degree days, cooling degree days.
Precipitation	Rain and snow loads affect building form, structure, and site formation. Precipitation patterns must be dealt with or can be used to enhance performance and comfort. Reflection from snow or frequency of rain storms directly affect solar incidence on the building. Storm direction, velocity, and ice build-up or snow accumulation important.	Seasonal and daily pattern from local and national data. On-site measurement. Observation of rain runoff or snow and ice build-up.
Humidity	May require addition or reduction for comfort. Evaporative cooling, fountains, or ponds for addition. Coolth tubes, dessicant absorption, dehumidification devices for load reduction. Affects choice of fabric of structure.	On-site monitoring, local and national records of seasonal and daily pattern (hourly).
Weather patterns	Length and frequency of combined climatic factors (wind, rain, snow, humidity, fog, sun, tornados, hurricanes) affect all aspects of solar and structural design. Building should metabolize with all normal patterns.	Local and national records, input of neighbors, farmers, builders, and code officials.

Issue	Effect/Need	Evaluation Method
Building Factors		
Construction techniques	Local building habits to include special skills or lack of them. Labor force availability, equipment required, trades and unions all affect design, building methods, construction season, working drawings, structural engineering, etc. Determine suitability of any design approach. Each geographic region has its particular reasoning.	Local contractors, local observation, code officials can all lend pertinent information.
Materials	Availability of on-site and commercial materials affects structure and fabric of building, heat sink, cost, design approach, etc. Critical to feasibility of and make-up of project. Limits of availability.	Local contractors and suppliers, site and local observation, owner research availability.
Costs	Local economics vary widely. Materials and labor costs are critical to the choice of materials, construction technique, and timing of project. Availability of money as well as building cost can affect size, quality, and design approach.	Research material prices at local suppliers. National and regional cost data services. Contact owners and contractors of recent construction. Real estate agents are a barometer of architectural values.
Planning Factors		
Codes/ordinances	City, county, and state code standards are legal requirements to protect health, safety, and welfare. Local ordinances can affect any or all building elements. Often conventional systems will be required even if alternative systems such as solar or waste treatment are allowed.	Obtain copies of all adopted codes as well as local design restrictions and code requirements. Contact local code officials for code checklist or compliance forms. Check with local contractors regarding compliance "techniques."
Zoning	City and county zoning requirements will affect development density, building placement, side yard setbacks, drive and road location, structure height, building type and use.	Research all local zoning information and requirements for the site and adjacent areas. City, county, and subdivision agencies.
Vernacular/historical	Often existing architecture will play an important role in determining the style of design. Clients or community may require specific styling for esthetic reasons. In many cases, vernacular architecture incorporates design features that will naturally enhance building performance and comfort.	Observation and research of local or regional architectural style ordinances. Talk to the "old timers" to find out what was and what worked!
Lifestyle	Microclimates and regional factors often influence esthetic, functional, and comfort standards. It is important to understand what people in an area react to for reasons of emotional and physical comfort, and why.	Research local architecture. Real estate agents are a valuable resource for determining desirability and the reasoning. Home tours are a good research method for residential ideas.
Utilities	The availability, cost, and type of utilities will affect the location, design, and operation cost of a building. On-site energy use vs. power grid connection, and sewage hook-up vs. on-site disposal or recycling are determined by utility realities.	Consult local energy, water, T.V., telephone, waste disposal, and sewage companies and agencies for rates, availability, and access. Alternative schemes should be evaluated by contacting local supply, installation and service dealers.
Access	Access to and within a site may be determined by ordinance, topography, view, density, etc. This is a primary factor for site planning, building location, and approach.	Site observation, right-of-way easements, traffic count and flows, local materials. Both pedestrian and vehicular patterns should be researched.
Adjacent use	Existing and future land use of adjacent properties and neighborhoods will influence solar access, view, general environmental quality of the design, and use of any building.	Local observation, planning maps and agencies, real estate agents, "the neighbors."

air quality or freshness

occupant acclimatization to geographic and climatic factors

occupant age, gender, weight, activity, clothing

interior space characteristics, including color, texture, volume.

The first four items on this list are directly related to the thermal environment and can be manipulated so that the body will seek metabolic thermal equilibrium, heat loss will be adjusted to the type of activity and heat available within a space, and the occupant of the space will feel comfortable.

Assuming an approximate relative humidity of 50 percent, a mean radiant temperature of about 70°F (21°C), and low air motion to be fixed factors, comfort is experienced when the air temperature is in the general range of 60–80°F (15.5–26.5°C). In fact, a variation of air temperature, which is common in passive solar buildings, is healthier than one daily or seasonal fixed value. In light of recent energy-conservation standards, recommended winter interior air temperatures have been reduced from 77°F (25°C) set in 1965 to 72°F (22°C) in 1975, while the summer value is set at 78°F (25.5°C). These have been recommended in the American Society of Heating, Refrigerating and Air Conditioning Engineers (ASHRAE) Standard 90–75. Suggested winter zone temperature values in the mid-60°F (about 18°C) range are not uncommon, especially in the kitchens and the bedrooms.

Another ASHRAE recommendation was a reduction in the level of relative humidity during winter from 70 percent in 1965 to 30 percent in 1975. This is energy-conserving, since more energy is required to maintain a 70 percent value in most climates. Within the range of 20–60 percent, relative humidity has only a small effect on comfort. When high humidity is a problem, however, air motion becomes an increasingly important factor because it facilitates evaporation. Additionally, lack of air motion often causes people to complain about the air's being stagnant. Gentle air movement results in a feeling of freshness and helps equalize temperatures within a space.

An often overlooked but important factor, especially with passive solar buildings, is mean radiant temperature. This is the weighted average temperature of the various radiant surfaces within a space at a particular point. Although the MRT tends to stabilize at the room air temperature, it is greatly affected by such factors as large amounts of glass, lights, and thermal mass (heat-storage capability). In a space with substantial thermal mass at 70°F (21°C) during winter, air temperatures can be in the low 60°F (15.5°C) range without a significant reduction of comfort. Such heat storage also allows a thermal lag, or flywheel effect, to occur. For example, if the air temperature in the space is reduced to 50°F (10°C) for a week or so, the MRT would slowly approach this lower value and keep the space comfortable for a longer period than would an identical space with no thermal mass.

Other factors affecting comfort are more dependent on individual physical and psychological responses. Although colors used in a space have little or no quantitative thermal significance, they can, and often do, affect the way a person perceives comfort. Proper color selection can reduce the amount of energy needed for space conditioning; during summer, greens and blues encourage cool sensations, while red, orange, and yellow hues can be used during heating periods to create illusions of warmth.

Another factor is acclimatization within a geographical location—that is, the tendency of people to perceive identical climatic situations differently. People adapt to the climate they

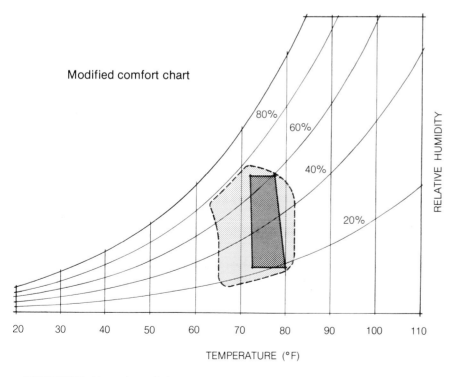

Modified comfort chart

RELATIVE HUMIDITY

80%

60%

40%

20%

TEMPERATURE (°F)

Normal comfort zone
Expanded comfort zone

live in, probably in order to reduce the strain on their metabolic furnace in a given climate over a long period of time. Thus, a 60°F (15.5°C) air temperature that may be pleasant and even warm to an Alaskan might be chilling to a Florida resident.

These and other items can and do affect the way people perceive, use, and relate to their environments. Careful balancing of all the comfort parameters can subtly or significantly modify environmental conditions, and each is important as a tool for maintaining the inhabitant's comfort.

Energy-Conscious and Passive Solar Design

Understanding the climate at each level is only one requirement for successful passive solar design. Also fundamental is an integration of common-sense design principles and the application of energy-conserving measures.

Energy-Conscious Design

All homes built today should attempt to conserve and save energy as an essential ingredient of common-sense design. A cold-climate energy-conserving house, for example, should exhibit such construction characteristics as a thermally tight weatherskin incorporating adequate insulation, double or triple glazing where appropriate, and high-quality weather stripping. And this is only a small sampling of items that will assure a successful first step in the direction of conservation.

It is essential that designers consider each energy-saving element within the overall design concept. Involving additional energy-conservation measures increases the probability of designing energy-conscious buildings. Energy-conscious opportunities are possible in all phases of building planning, design, construction, and maintenance. The importance of these op-

portunities varies with the type and scale of a project, but all should be considered. Macro-, micro-, and interior climate characteristics, when properly interpreted and utilized, become energy-conscious design guides.

The building's configuration or shape should take into account the surface-to-volume ratio. The greater the wall and roof areas of a building, the higher the ratio, and the greater the potential for heat loss and gain; this may be an advantage in mild climates, but it is certainly a disadvantage in severe climates. Optimum building configuration (length to width to height) must be evaluated against such factors as wind flow, internal ventilation, function, and solar exposure.

Building surfaces should be suited to a particular orientation. North elevations or facades should be different than south, west different than east. Often ignored in conventional design but obvious to good passive design is solar orientation. Understanding where the sun is at different times of the year and how the building conforms to its impact is essential. Another element to consider is the wind direction. Prevailing winds can be caught for summer ventilation, and, just as important, one should buffer against winter winds to reduce heat loss. Locating entries away from icy, snowy, and windy exposures is particularly important in northern climates to facilitate entrance and exit as well as reduce infiltration losses.

Adequate and proper insulation is a primary energy-conservation measure. Energy-conscious design requires evaluation of comparative techniques and materials for proper insulation, and their appropriate location or placement. For example, highly insulative urethane foams should not be used below grade, since they absorb moisture; instead, a lower-value,

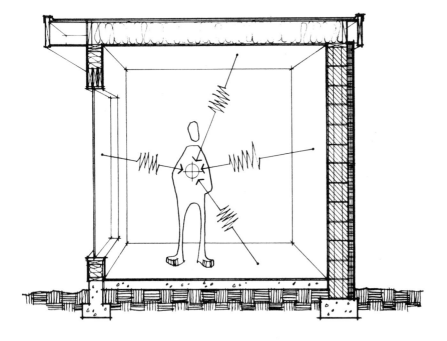

Mean radiant temperature (MRT) is the weighted average of the various radiant surfaces within a space.

closed-cell, extruded, water-resistant polystyrene is often more suitable and effective. Other building-envelope determinants include the selection of proper finishing materials, giving consideration to color, texture, durability, and maintenance, while examining response to sun, winds, and moisture. High or low light-absorptive materials and finishes should be carefully chosen to absorb or reflect heat gain as climate, orientation, and energy demands warrant. A dark-colored roof may minimize winter heating requirements, but the added summer cooling load may more than offset this asset. To reduce mildew and rot, control of moisture and condensation within walls or roofs or at their surfaces is also quite important, especially in the tight construction of high-insulation standards.

Windows are prime elements affecting the control of heat loss and gain. Glazed openings should be designed to accept or reject solar radiation according to orientation, time of day, season, and the inhabitant's needs. Proper sizing, locations, and types of windows encourage daylight. Like operable windows, openings such as doors, louvers, vents, turbines, and exhausts all facilitate ventilation. Double or triple glazing and the use of thermal shutters minimize heat transfer under extreme weather conditions. Window frames that form direct thermal links between the interior and exterior should be avoided. Finally, control of light and heat gain can be accomplished with screens, overhangs, and shutters on the exterior, and drapes, blinds, curtains, and shutters on the interior.

Interior design decisions made at both the planning and installation stages strongly influence the energy needed for space conditioning. The zoning of interior spaces can create areas which act as buffers against external walls and harsh climate. Corridors, utility spaces, closets, service rooms, and other low-use areas do not require close temperature control and make excellent buffer spaces. Air-lock entries are another example of buffer zones appropriate to extreme climates. Open-space planning and multiuse of spaces are becoming more common in energy-conscious homes. Open-space planning allows more even heat distribution and ventilation within an area as well as providing the opportunity for one spatial function to flow into another; multiuse of space can save energy by reducing the square footage requirements of a building by eliminating spaces designed for only one use.

Selection and sizing of conventional heating and cooling equipment should be done carefully. Systems which capitalize on waste-heat recovery and provide automatic setbacks and temperature controls should be used. With passive solar systems, particular attention should be given to the integration of the auxiliary system. The auxiliary or backup can be significantly minimized when it is radiatively or convectively coupled with the operation of the passive design. This is because the thermal storage inherent in passive design creates a thermal flywheel effect, capturing and holding both solar heat and much of the heat supplied by the auxiliary source and slowly releasing it over a long period of time.

Finally, and not to be forgotten, are the materials used. These are often overlooked as a means of conserving energy. The most effective material for a particular span and load should be selected. A wood truss, for example, requires less energy to produce than a steel girder. Energy used in the production, transportation, and maintenance of architectural systems should be reviewed from a life cycle energy-use standpoint, just as components in appliances are examined

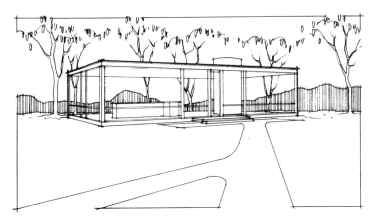
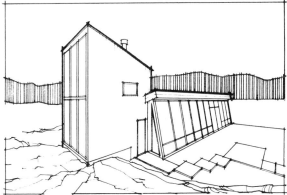

The nonoriented building at left differs from the energy-conscious building at right, whose form is influenced by sun, wind, and function.

for their overall efficiency. Concrete block requires more energy to produce than wood or adobe; polyurethane is a much more energy-intensive insulating material than pumice or cellulose. With spiraling energy costs, materials will soon be priced according to their total energy worth, not just availability.

Energy-conscious design options are the first line of defense in passive solar buildings. Conserving energy is generally the most cost-effective approach to reducing a building's energy load. By its proper use, a designer is given greater latitude in incorporating the passive system.

Passive Solar Design

Efficient operation of passive solar systems is achieved by three basic heat-transfer processes: conduction, convection, and radiation. Since pure passive systems use only natural energy flows, some user involvement is required to direct the flows both within a building's spaces and at its weatherskin. Small fans and pumps are sometimes used to adjust or augment a particular aspect of the design.

Passive solar heating can be divided into the following three generic categories:

Direct gain—heat is collected and stored directly in the living space
Indirect gain—heat is collected and stored adjacent to the primary living spaces and thermally linked to them
Isolated gain—heat is collected adjacent to or apart from the weatherskin and stored either apart from or in the living spaces

The primary elements to consider with each category include the following four aspects of collection, storage, distribution, and control:

Solar collection surfaces are generally transparent or translucent glass, plastic, or fiber glass, oriented toward the sun. When selecting materials, pay careful attention to material degradation by solar exposure and other weathering elements.
Thermal storage materials include concrete, brick, sand, tile, stone, and water or other liquids, as well as phase-change materials (those which absorb or release heat when changing from solid to liquid, such as eutectic salts and paraffins). Storage units should be well placed to receive proper solar exposure, either directly (radiatively) or indirectly (convectively). Materials with adequate thermal storage capacity absorb and retain the sun's heat until it is needed, minimizing daily internal temperature fluctuations.
Heat distribution occurs through natural means by conduction, convection, and radiation. Fans or other mechanical equipment is sometimes used to improve comfort and increase system efficiency. They may be helpful in regulating heat flows and avoiding over- or underheating of particular spaces.
Control mechanisms such as vents,

This water wall (glazing, water drums, overhead fan, and exterior shades) illustrates the four primary elements of passive heating.

dampers, movable insulation, shading devices, and other aids are helpful to encourage and ensure a balanced heat distribution. When installed at crucial locations, these aids release, channel, trap, or block thermal energy.

Passive solar cooling, like passive solar heating, is simply the tempering of interior spaces by optimizing the use of natural thermal phenomena. A structure designed for natural cooling ideally incorporates features that minimize heat gain. When possible, external heat gain should be controlled before it reaches or penetrates the weatherskin. Also, internal heat gain by lights, people, and equipment should be reduced, by both design and management.

Dissipation of heat is accomplished by cooling the interior mass, the air, or both. As in passive solar heating, conduction, convection, and radiation are the thermal transfer processes used. Additionally, evaporation and dehumidification are valuable aids. Many passive cooling methods exist: cross-ventilation, radiation to the sky, day-to-night closing and opening of walls or roofs, induction of precooled air, night cooling of interior air and building mass, earth tempering, water evaporation, and desiccant masses, among others. This science integrates traditional architectural solutions, modern materials, refined knowledge of thermal dynamics, and use of all of nature's helpful patterns. The following are the three generic cooling categories:

Direct loss—heat is dissipated directly from the space, as in cross-ventilation

Indirect loss—heat is dissipated at the weatherskin or through intermediate elements, such as cooling of mass walls

Isolated loss—heat is dissipated away from the weatherskin; for example, induced air can be precooled in the earth's mass

Passive solar domestic water heaters are often integrated with passive building designs. Unlike active solar water heaters, passive systems do not require pumps or mechanical devices to capture or store the sun's heat. Instead, they rely on natural convection, as in thermosiphoning, or preheating of the water supply before entering a conventional water heater, commonly referred to as a batch or preheat system.

As research and development progresses, photovoltaics will continue to reduce in initial cost and increase in life and efficiency. Direct conversion of the sun's energy into electricity for powering lights, appliances, and equipment can turn a passively heated and cooled house into one that is totally energized passively.

The specific tasks of heating domestic water and supplying electric power for a building can often be accomplished by passive on-site means. Localized power from wind, water, the earth's heat, the sun, hydrogen production, methane generation, and other potential sources are becoming feasible realities. Modern needs and technologies are making these types of solar energy today's resources. These myriad aspects of solar design are decidedly cost-effective in many cases and should be considered in all future designs.

Planning and Design Decision Making

How are the elements of climate analysis, energy-conscious planning, and passive solar design molded into the most suitable building solution? What are the most cost-efficient energy features to incorporate? What passive solar system or systems are best for a particular location and program? These are only a few of many questions that are addressed through the process of planning and design decision making. No one solution is necessarily best under a given set of circumstances. For instance, direct-gain buildings may generally be more efficient, but their potential for localized interior overheating and glare is greater. On the other hand, indirect-gain systems may minimize localized overheating, but they could create excessive nocturnal thermal losses through the glazing if they are not designed properly.

When a building is designed, purely objective evaluation sometimes overshadows the owner's preconceptions. In a building that is to flow with nature, the esthetic experience is as important as the thermal performance. The successful combination of art and engineering creates an environment that is both exciting and satisfying to the inhabitant. This is the essence of the beauty and logic in passive solar architecture.

Logical passive solar design requires a quantitative assessment of the building's thermal energy flows. However, the performance of a passive solar building in a particular climate is somewhat difficult to predict with a high degree of accuracy. Improper assumptions during design can lead to undue discomfort and costly alterations. A passive building that might perform quite well in the winter may overheat in the summer; conversely, a passive building may be cool and comfortable in the summer, but may underheat in winter. It is desirable, therefore, to evolve designs which function efficiently at critical seasonal points, and provide proper levels of thermal comfort to the occupants throughout all seasons.

There are various technical methods and tools available to the building designer for use at different stages of the design process. During the planning and conceptualization stages, reference manuals assessing climate and general design guidelines are helpful (see Bibliography). With these planning resources, a designer can begin to gain a good understanding of a particular climatic area. Additionally, suggestions are made to overcome various climatic liabilities and to augment climatic assets. For example, Washington, D.C. has outside mean temperatures less than 50°F (10°C) for 37 percent of the year. To combat this climatic liability, compare the following design principles and their overall effectiveness, as computed by Donald Watson and K. Labs in *Climatic Design for Home Building* (Guilford, Connecticut, 1980):

Minimize infiltration   76% effective
Minimize conductive
heat flow   69% effective
Promote solar gain   66% effective
Minimize external
air flow   66% effective

At the building-design stage, generalized rules of thumb can provide easy-to-follow preliminary sizing instructions. Examples of these methods are described in several publications (see Bibliography). These books describe manual calculations which vary with respect to the time required to use them. As more precise flexibility or suitability to a particular location increases, so does the complexity of the process and application.

More detailed and accurate sizing methods appropriate to design refinement require more time and technical experience. Hand-held calculators, those economical modern wonders, can be programmed to handle some of the more time-consuming simple calculations. Programs such as PASCALC II (by Total Environmental Action, Inc.) and PEGFIX (from Princeton Energy Group), among others, offer a simplified analytic method that substantially reduces the time required for other, more sophisticated thermal analyses. The main advantage of this design tool is its ability to handle more input data faster and with a higher degree of accuracy than some manual methods or rules of thumb. They are economical enough to be cost-justified by an individual or a small office.

Rules of thumb and simplified manual or programmed design methods provide insight and guidance for the preliminary design, including proportioning, orientation, weatherskin, glazing, and thermal mass. However, these methods cannot fully optimize seasonal performance, since these simplified design procedures are generally derived from extensive computer modeling of simple solar buildings upon which a large number of generalized assumptions have been made. For the final design stages, sophisticated computer programs offer a higher level of analysis, since they are capable of taking into account more varied conditions, such as building geometry, movable insulation schedules, absorption and conduction of solar radiation through opaque surfaces, nonsouth window gain, shading devices, hour-by-hour weather data, and many other factors. These methods are desirable in the development and working-drawing stages to suit the building precisely to its climate. Just as a car is tuned for optimum mileage, a building can also be tuned for optimum performance under given operating conditions.

Computers range in size, cost, and capability from micro- to mini- to large mainframes. Both mini- and mainframe computers can use programs capable of processing building types ranging from simple residential to very large and complicated commercial buildings. Programs such as CALPAS3, DEROB, DOE-2, FREHEAT, SUNCAT, and others are currently used on mini- and mainframe computers. At the present time, there are relatively few simulation programs available to professionals for use on microcomputers except on a time-sharing basis (where a small microcomputer or terminal ties into a mini- or mainframe computer). Due to the complex nature of many of these programs (which allows for their greater flexibility), the time required to enter all the required data can be somewhat extensive.

In selecting the design tool for analyzing the thermal performance of a number of the projects featured in this book, a balance between input time and accuracy of results was desired. The CALPAS3 computer program was chosen to meet our objectives. The original CALPAS program was developed at California Polytechnic State University at San Luis Obispo, California, by Professor Phil Niles and collaborators; CALPAS3, an enhanced version developed by the Berkeley Solar Group of Berkeley, California, handles most of the generic types of passive systems and can give a detailed hourly readout of temperatures and energy flows, in addition to monthly and yearly energy flows and temperatures. This program requires a fairly detailed building description as input and is run on a main-frame computer. However, the time required to enter information and run the program is relatively short compared to involved manual or hand-held calculators and other complicated programs run on the larger computers. Programs similar to CALPAS3 are usually available through specialized solar consultants. Appendix A provides a description, and explanation of a typical input and modeling procedure actually required for the thermal analysis used for the projects in this book.

Performance simulation is an invaluable aid to the designer, both as a design tool and as an energy-efficiency rating of passive solar buildings. With it, the optimum "mix" of energy-conscious and passive-design strategies can be evaluated. By simulation, the sizing of passive solar elements developed during the architectural design process can be refined to maximum effectiveness. This method of calculation reinforces, validates, and quantifies intuitive decisions, a creative force not to be eliminated. In this case, the beauty is justified by logic.

Chapter 2

DIRECT SYSTEMS

Direct systems are the most commonly used in passive solar buildings, and are generally the most efficient way to use the sun's heat. Since most buildings have windows facing the sun, they utilize direct gain to some degree. Simply stated, in a direct-gain system, heat is collected or dissipated directly within the living space; thermal collection and storage are integral with the building's interior. Solar-oriented windows (collectors) admit winter radiation, and the interior spaces contain adequate amounts of thermal storage materials incorporated in the building structure, or furnishings which absorb solar energy. During the cooling season, windows, walls, and roofs are opened for natural or induced ventilation, directly cooling both the thermal mass and interior space. Fans and other mechanical devices are not critical to the system's operation or comfort.

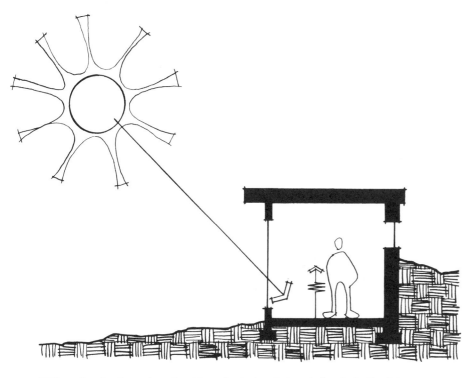

With a direct-gain system, heat is collected and stored directly in the living space.

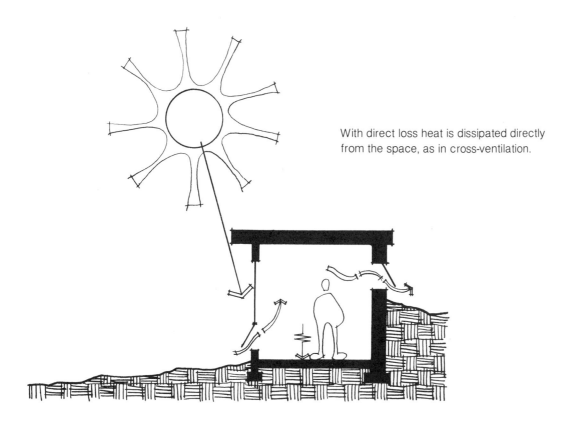

With direct loss heat is dissipated directly from the space, as in cross-ventilation.

This functional and sculptural shape has strong geometric lines and natural exterior materials, texture, and color. The central tower provides vertical circulation and air flow. Sayre residence (Sunburst), Applegate, California.

Functional geometry: the forms all serve solar, as well as architectural, purposes. Sayre residence.

Nestled in the foothills of the west slope of the Sierra Nevada, this home was systematically designed to consider the microclimate. The climate in winter is cool and often wet; snowfall and subfreezing temperatures are not uncommon. In summer, daytime temperatures often exceed 100°F (37.5°C), the humidity is low, and night temperatures regularly drop into the mid-60s. With fairly high utility rates and the need for space conditioning throughout the year, passive solar design was the most appropriate architectural solution.

The client's request was for a strong architectural statement exemplifying both modern design and passive solar features. The solution, a rational and funtional sculpture, has bold geometric lines, natural exterior texture and colors, and a clean, bright interior. Its volumetric forms are startling, yet blend with the environment. The architectural elements of this simple earth-integrated, three-zoned residence are suited to a particular passive solar function, thus creating a play of architectural use against natural energy flows and their effects.

The geometrics are planned to shade from and open up to the sun at required times. The owners manage to "sail" their home with a minimum of manual operational steps.

Downstairs, double glazing and the thermal storage masses—the elegant tile-on-concrete slab floor and the massive white concrete masonry walls—trap and store heat from the low winter sun. An energy-efficient fireplace with glass doors, outside combustion air, and heat-recirculating features provides auxiliary heating. The master bedroom is heated by direct gain and rising warm air from the lower main level.

The central tower is also functional, used for vertical circulation and thermal air flow. An operable skylight at the top allows warm air to escape from the highest point of the building, which in turn induces outside cool air through any open doors, windows, or skylights. This process is known as the chimney effect, or stack ventilation.

The summer operation consists of opening the high- and low-vent windows both upstairs and down each evening, and closing them each morning. Vertical exterior sun shades are adjusted occasionally to minimize sunlight to the south glass, while the north wall is extensively bermed for additional cooling. This combination of elements maintains comfort in the warm seasons. Additional humidification and evaporative cooling can be achieved by irrigating the plants outside the low-air-intake windows.

A hot-water preheat tank provides approximately 70 percent of the domestic hot-water needs. Natural light and dimmer-operated low-energy task lighting provide all the necessary lighting.

We are particularly elated with the fact that this house has met or exceeded our expectations in terms of thermal performance and, most important, livability. Our logical design approach has yielded a beautiful home.

From the Inhabitant

After a year of living in the house, our hopes were more than met. First, some statistics we logged using simple recording thermometers during January and February:

	Inside	Outside
Average maximum	75°F (24°C)	60°F (15.5°C)
Average minimum	63°F (17°C)	41°F (5°C)
Peak maximum	81°F (27°C)	73°F (23°C)
Peak minimum	59°F (15°C)	31°F (−.5°C)

There were twenty-six sunny days in this relatively mild period. Fires were built twenty-two days, and the house was vented three days. The only other source of heat was a single baseboard heater used to warm the master bedroom in the mornings. Less than one-third of a cord of wood was burned, and the total household electrical use during the period was about 1800 kwh, or $1.10 a day. No insulated window coverings were used, as we have chosen to not install them.

Although we have not logged data for the summer, we recall that the only time the inside temperature exceeded 80°F (26.5°C) was during a prolonged July heat wave, when the outside

Earth integration helps to heat and cool the house. Sayre residence.

temperature never dropped below 70°F (21°C) and usually reached 105°F (40.5°C). With exterior window shades and an evaporative cooling assist from vegetation now being planted on the bank below the house, we expect the natural cooling system to work even better.

This impressive thermal performance was achieved with only a few minutes of our attention each day: building fires in the winter and operating windows in the summer. No adjustments in our schedules were needed, fortunately, since our jobs left us no time to tend the place. For this small effort, the house rewarded us with a modest winter utility bill.

We had planned to live in the Sierra Nevada foothills, and were inspired to build a passive solar home after attending local classes. We had always shared a preference for contemporary design and natural materials; living in five other homes typical of urban developments reinforced these preferences.

The house is a joy to the senses. Interior textures of wood, tile, and plaster complement the outdoor vista. The clean, contemporary simplicity of the exterior structure is in harmony with the natural setting. Our builder carried out the design with detailed, quality construction. We have named our home Sunburst in recognition of

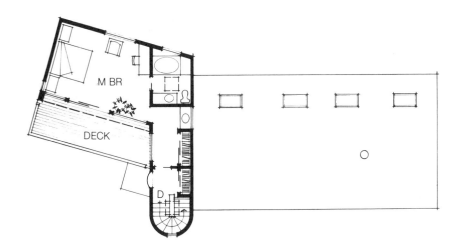

Upper level

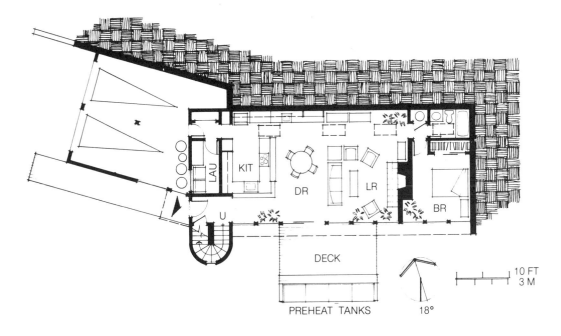

Main level

the level of inspiration in both its design and execution.

We enjoy the solar processes at work in this house, and like the way it enhances the environment. The passive system is silent, enabling us to hear birds, wind in the trees, or even distant coyotes. Lights and colors change hourly and daily as the seasons unfold.

Equally important to us is the house's livability. Its layout permits privacy and separation, is perfect for the two of us, and can still accommodate large gatherings of family and friends. The truth of Sunburst is its success in these basic areas and in its overall esthetic appeal. We take the passive system for granted as a natural part of our lives.

Clean, bright, and open interior accepts the winter sun. Sayre residence.

1. winter solar gain
2. radiant thermal storage mass
3. earth integration
4. energy-efficient fireplace
5. shading
6. ventilation
7. solar water heating

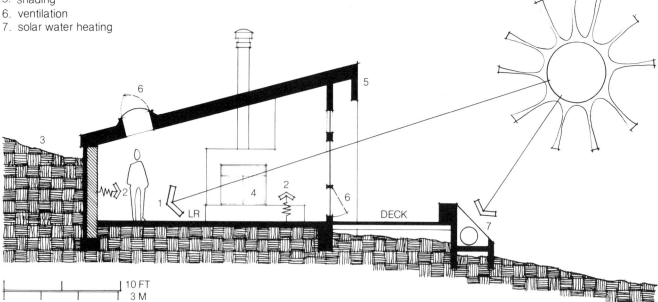

Section

Project Data Summary

Project Information

Project: Sayre residence
     Applegate, California
Architect: SEAgroup—David Wright, AIA, Dennis A. Andrejko, AIA
     Nevada City, California
Builder: Peoples Places: Tom Huiskens
     Auburn, California

Climate Data

Latitude	38.8°N
Elevation	1,900 FT (579m)
Heating degree days	3,105
Cooling degree days	1,079
Annual percent possible sunshine	70%
January percent possible sunshine	45%
January mean minimum outdoor air temperature	35°F (1.5°C)
January mean maximum outdoor air temperature	54°F (12°C)
July mean minimum outdoor air temperature	62°F (16.5°C)
July mean maximum outdoor air temperature	93°F (34°C)
Climate features: cold, wet winters; warm, dry summers	

Building and System Data

Heated floor area	1,650 FT^2 (153m^2)
Solar glazing area	
Direct gain	362 FT^2 (33.5 m^2)
Thermal storage capacity	
Tiled concrete floor	10,266 Btu/°F
Concrete masonry retaining wall	6,250 Btu/°F
Interior masonry walls	2,950 Btu/°F

Performance Data

Building load factor	7.3 Btu/DAY °F FT^2
Auxiliary energy (heating)	6.9 MMBtu/YR
Auxiliary energy (cooling)	1.1 MMBtu/YR
Solar heating fraction	80%
Night ventilation cooling fraction	97%

Simple, strong, and straightforward, this home is set into a hillside and nearly buried in the earth. Hadley residence, Minneapolis, Minnesota.

The key to the success of most passive solar architecture is simplicity. Adding unnecessary fans and exotic shades, or creating complex forms and intricate spatial relationships, often creates design liabilities with more difficulties than the problems they attempt to solve. (Less successful passive buildings are often characterized by system complexity, with devices working against the natural laws of thermal dynamics rather than with them.) Designer Tom Ellison has created an excellent example of passive simplicity with this Minnesota residence. Located on a steeply sloped, heavily wooded site in a cold climate, the house is set into the hillside with access from the south.

The plan is simple, strong, and straightforward. All the major living spaces open to the south and have excellent views and good daylight. The living/dining area is located in the center of the plan, with the children's and adult's bedrooms at opposite ends of the house. The building form is simple. A low sloping shed maximizes southern exposure and, with earth berming, virtually eliminates any exposed north wall. Except for two double-glazed skylights, all glass faces the sun.

The passive solar system is also simple. The direct-gain design uses double-height glazing, which maximizes solar penetration deep into the space during winter, heating the tile floors and interior masonry walls. Bold pivoting exterior shades and the high fixed overhang allow total seasonal shade control. The subtlety of the darkened exterior south wall adjacent to the windows increases total solar absorption at the weatherskin. The auxiliary heating system comprises a wood stove and an electric forced-air system with supply ducts below the floor slab. The return air ducts run continually along the peak of the ceiling to collect and redistribute warm stratified air. This serves as a preheat to the mechanical system, but often only the furnace blower is used to recirculate the warm air.

The synthesis of these simple elements produces a strong architectural solution. This house proves that one need not get exotic to provide high solar performance and a delightful space for the occupant. In evaluating cost-effectiveness and performance, here is a confident affirmation that less is more.

Bold, pivoting exterior shades and high fixed overhang allow summer sun control. Hadley residence.

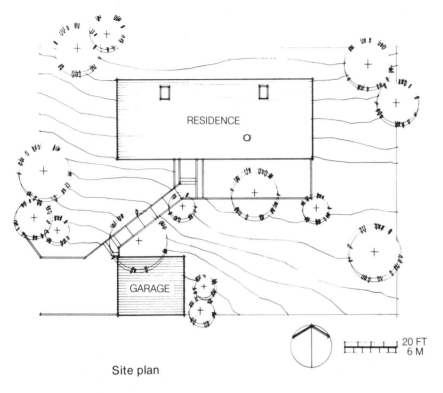

Site plan

From the Designer

There were four basic program requirements for this house:

1. Three bedrooms, two baths
2. Quality and excitement throughout (the owner stated at the beginning of the preliminary design phase that "the interior spaces should be flooded with light and have a bright, natural quality")
3. High energy efficiency
4. Moderate construction budget

"Simplest things first" characterizes the design method used to meet the above requirements. A serious effort was made to blend the most cost-effective aspects of passive solar design with extremely efficient conservation measures.

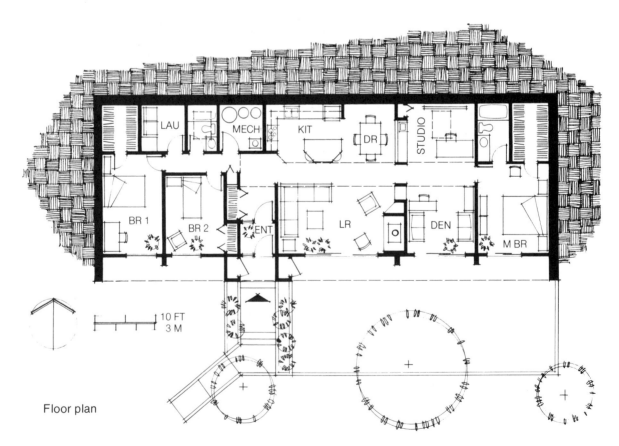

Floor plan

The conceptual key to this design is the coupling of a direct-gain system with earth sheltering. The large massive floor and concrete block walls, required to support the earth loads, are also the storage medium for the absorbed solar radiation. There is a very large area of massive material in this house which, even though a good portion of it is at the rear of the living spaces, can still serve as acceptable, indirectly coupled thermal storage. Most of the interior space is painted white to reflect and diffuse the radiation to cover the entire interior, making the large surface area of dark floor more efficient as a thermal storage mass.

In Minnesota, with drastic seasonal temperature extremes and frigid winter winds, earth sheltering is a natural choice. In this case, it was placed against the walls, where it is more effective than on the roof (the construction costs of placing twelve to eighteen inches of earth on the roof are great, as is good-quality waterproofing and rigid insulation). It should be noted as well that shallow earth on the roof has no real significant effect on heat loss and, therefore, the more cost-effective approach of "super insulation" was used. Interestingly, the surface area of the roof accounts for 50 percent of the total exterior surface area of the house, but only 7 percent of the total building heat loss. The earth profile on the east and west sides is also the result of cost tradeoffs: by sloping the berm down to meet natural grade on the south side, no retaining walls were required.

Earth berming virtually eliminates any exposed wall. Hadley residence.

Even though most of the design concepts addressed energy and cost, much effort was placed on blending these criteria with a stimulating and esthetically pleasing environment. A key to this was limiting the design to an extremely simple exterior shell and maintaining the beautifully wooded natural setting. The interior was kept very open, with all spaces having a clear view of the woods. In the living areas, the long indirect light bridge provides some lowered ceiling areas for variety, a very natural indirect lighting effect in the evening, and a great place for plants.

1. winter solar gain
2. radiant thermal storage mass
3. earth integration
4. energy-efficient wood stove
5. air ducts
6. shading
7. pivoting shade
8. deciduous trees
9. suspended ceiling bridge

Tile floors and brick walls store winter heat. Hadley residence.

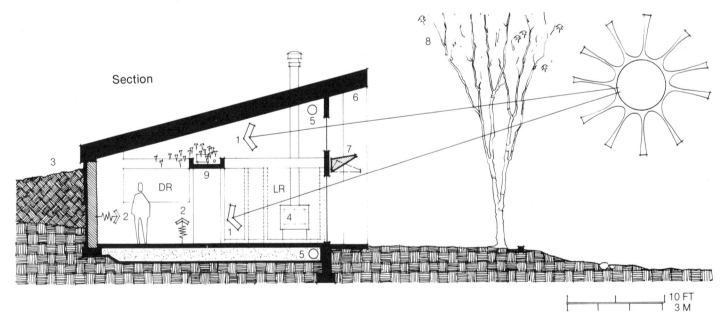

Section

Project Data Summary

Project Information

Project: Hadley residence
          Minneapolis, Minnesota
Designer/Builder: Tom Ellison
          Minneapolis, Minnesota

Climate Data

Latitude	44.8 °N
Elevation	830 FT (253m)
Heating degree days	8,382
Cooling degree days	578
Annual percent possible sunshine	58%
January percent possible sunshine	50%
January mean minimum outdoor air temperature	3°F (−16°C)
January mean maximum outdoor air temperature	21°F (−6°C)
July mean minimum outdoor air temperature	61°F (16°C)
July mean maximum outdoor air temperature	82°F (27.5°C)

Climate features: long, cold winters; short, hot summers

Building and System Data

Heated floor area	1,955 FT^2 (181m^2)
Solar glazing area	
Direct gain	350 FT^2 (32.5m^2)
Thermal storage heat capacity	
Tiled concrete floor	17,760 Btu/°F
Masonry walls	33,600 Btu/°F

Performance Data

Building load factor	5.3 Btu/DAY°F FT^2
Auxiliary energy (heating)	44.9 MMBtu/YR
Auxiliary energy (cooling)	0.5 MMBtu/YR
Solar heating fraction	63%
Night ventilation cooling fraction	91%

The house reflects traditional New England designs. Holdridge residence, Hinesburg, Vermont.

In Vermont, people tend to build long-lasting things that work well. The long winters with numbing cold and winds put passive solar designs to the test, and comfortable winter living takes a little extra planning. Architect Robert Holdridge and his partner, Doug Taff, have experience designing passive solar buildings in this tough New England environment. One of their goals is to create a "$100-a-year-house": one that works with the microclimate and has an annual heating cost of $100 or less.

This design solution is a well-insulated, articulated box with south-facing glazing, adjustable window blinds, and plenty of internal thermal mass in the form of brick floors and walls. A striking feature is the dominant red brick and masonry spine that runs the length of the house. This beautiful form separates north from south spaces and is ideally placed to absorb internally generated heat from the kitchen, fireplace, and other sources. This simple approach is economical and very effective even in a severe climate zone. The authors share Taff's opinion: "From our experience with other buildings that we have designed, direct-gain buildings are well suited to the Northeast because they are pure and simple. Trombe walls, water walls, rock beds, and other solar systems are fine, but direct gain is the winner in Vermont housing."

The red brick and masonry spine runs the length of the house. Holdridge residence.

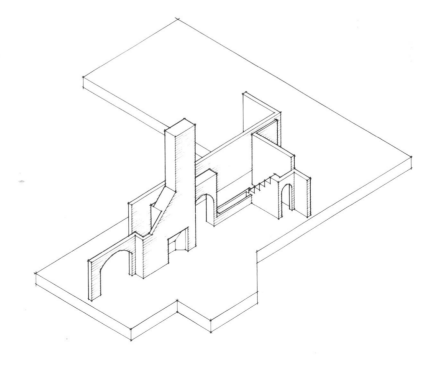

Brick and masonry thermal spine

The window areas of this stylish exterior allow natural light penetration to all areas. Holdridge residence.

From the Architect

This design incorporates the ideas of four years of residential design work employing passive solar techniques in one of the cloudiest and coldest climates in the continental United States. It is a testament to the power, predictability, and potential of solar architecture for this region.

Located in an area where the architecture is dominated by two-story, compact structures with minimum window area, this residence affords the luxury of open expansive living, a flow of inside to outside spaces, and abundant light. Because of the rural setting, privacy was not affected by large areas of south-facing windows.

The home was designed for a family of five. Time-of-day living considerations were met by separating the living and sleeping functions into distinct areas with different thermal characteristics. By dividing the house this way, daytime living occurs in the solar-gain portion while the bedroom section can remain cooler.

Our experience has shown that solar energy effectively provides heating, lighting, and cooling. This home employs all three functions. South-facing windows allow sunlight to pass into the interior; the solar energy strikes the masonry walls and floor, heating the structure. All floors and walls exposed to the sun are designed for heat storage. Window areas allow natural light penetration to all spaces; electric lights are only used at night. Cooling is accomplished by induced natural ventilation: high operable windows in the living section allow warm air to escape during the summer months. Fresh air is drawn and channeled through the entire structure simply by opening lower windows. The seasonal path of the sun, working with the large amount of mass within the structure, provides comfort throughout the year. Wood, a site-available natural resource, is used as the backup energy source for space heating and domestic hot water. This Vermont residence is totally free of energy inflation.

Passive solar architecture is a beautiful opportunity for a designer to bring a building's natural thermal dynamics into play. Major consideration was given to the design of the thermal storage units of this house. The heat is not stored in rock beds, water tubes, or water drums, but in the brick mass which is the architectural fabric that holds the spaces together.

From the Inhabitant

The predominant feeling produced by living in a direct-gain solar house is the awareness of unity with nature, both inside and out. The house is in rhythm with nature. The interior, while always comfortable, varies with the time of day, the weather, and the seasons. Energy independence and

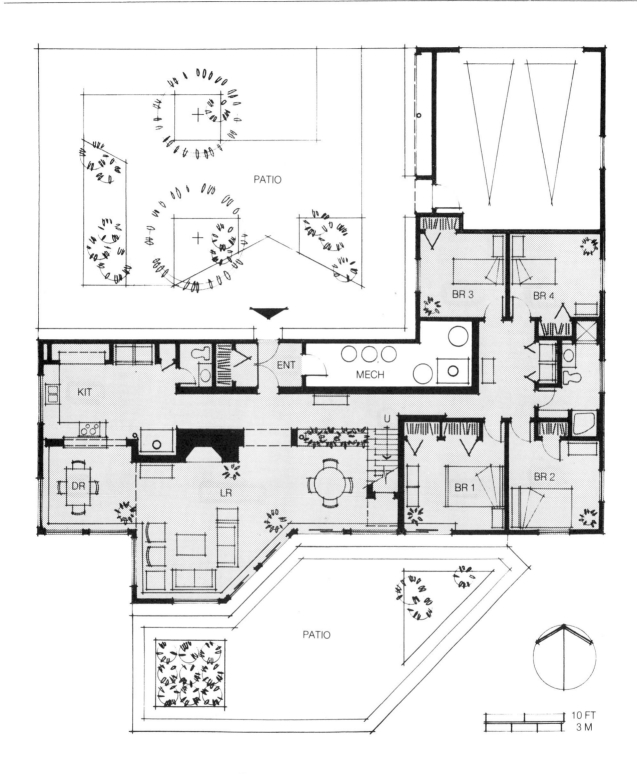

PATIO

BR 3

BR 4

KIT

ENT

MECH

DR

LR

BR 1

BR 2

U

PATIO

10 FT
3 M

Floor plan

The red brick spine separates north spaces from south and absorbs internally generated heat. Holdridge residence.

use of renewable resources provide a sense of security and stability to our family. The power of natural light was something I underestimated. The control of glare is especially important in a direct gain space; we installed Venetian blinds to direct the natural light.

One of the most pleasant surprises has been the health of our family and plants. Congested sinuses, which appeared like clockwork when the oil furnace was turned on for the winter in our previous home, have all but disappeared. And our Christmas cactus was in full bloom for Thanksgiving—an unexpected bonus!

1. winter solar gain
2. radiant thermal storage area
3. sun-control blinds
4. energy-efficient wood stove
5. solar water heater

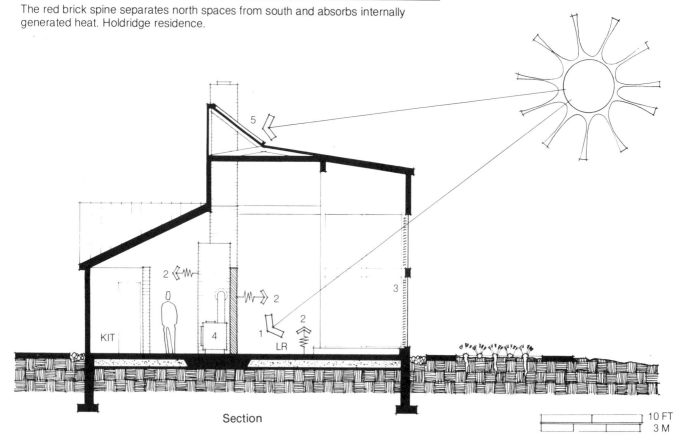

Section

10 FT
3 M

Project Data Summary

Project Information

Project: Holdridge residence
          Hinesburg, Vermont
Architect: Parallax, Inc.—Robert Holdridge
            Hinesburg, Vermont
Builder: Conrad Bosier
          Hinesburg, Vermont

Climate Data

Latitude	44.0°N
Elevation	540 FT (164.5m)
Heating degree days	7,721
Cooling degree days	396
Annual percent possible sunshine	46%
January percent possible sunshine	34%
January mean minimum outdoor air temperature	8°F (−13°C)
January mean maximum outdoor air temperature	26°F (−3°C)
July mean minimum outdoor air temperature	59°F (15°C)
July mean maximum outdoor air temperature	81°F (27°C)

Climate features: biting cold, long winters; short, warm summers

Building and System Data

Heated floor area	2,385 FT^2 (221.5m^2)
Solar glazing area	
Direct gain	623 FT^2 (58m^2)
Thermal storage heat capacity	
Masonry walls	78,790 Btu/°F
Concrete and brick floor	26,700 Btu/°F

Performance Data

Building load factor	4.6 Btu/DAY °F FT^2
Auxiliary energy (heating)	48.3 MMBtu/YR
Auxiliary energy (cooling)	0.3 MMBtu/YR
Solar heating fraction	60%
Night ventilation cooling fraction	99%

On the formidable Maine coast, Yankee ingenuity and solar energy combine.
Sunkissed, York, Maine.

Shed roof and shingled exterior reflect traditional building shapes of the area.
Sunkissed.

Architect Donald Watson was concerned with architectural alternatives even before serving as a Peace Corps volunteer in North Africa during the early 1960s. He has been actively involved with solar architecture for over a decade, and has written and edited several books and other publications on the subject.

This sun-kissed beauty, located on the Maine coast, is an example of a regionally suited, well-proportioned and -detailed passive solar design. It demonstrates a refined application of the elemental passive solar features, including proportion and placement of solar-gain windows on the south facade; sizing and location of thermal storage mass within the structure; and detailing and integration of insulative window shades. Each of these primary elements is engineered for optimum performance and is carefully composed as one part of the architectural whole.

The traditional masonry fireplace that is generally located near the center of New England homes is updated by allowing the sun to warm this vertical thermal storage mass directly and by replacing the colonial-style fireplace with an efficient cast-iron wood-burning stove. This central heat-

Upper level

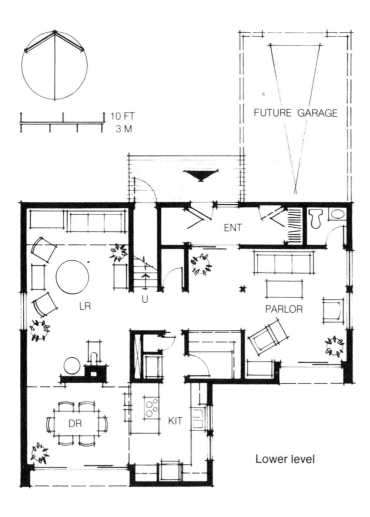

Lower level

ing system consists of several tons of masonry to act as a large interior thermal storage mass or thermal flywheel that changes temperature slowly and is warmed by both solar and wood energy.

Solar gain not absorbed and stored by the thermal mass is drawn off from high in the house by a thermostatically controlled fan to rock-bed storage located below the floor slab. This system prevents overheating of the interior when the sun is shining and stores the excess heat, which is then used in lieu of the auxiliary furnace or stove. Strategically placed interior openings allow convection and distribution of interior heat. Insulative shades are concealed in overhead window valances when open, and a tight side track minimizes convective losses around the edges when the shades are down.

Sunkissed stands out as a pragmatic and innovative example of Yankee ingenuity applied to passive solar architecture. Don Watson sums up his approach to energy conservation: "Architects once studied the rules of proportion for the styles and orders of the classic temple of antiquity. The earth is now that temple: the rules are those of building and living within the limits of the world's balance of resources and energy."

From the Architect

The design was conceived as a direct-gain system, with specific design features intended to overcome some of the traditional problems associated with direct gain: most of the glazing is located high on the south walls to minimize direct glare and radiation at the living level. This high clerestory also provides for temperature stratification, allowing the overheated air to rise to the top of the house. Thermal storage is provided in a masonry wall between the dining and living areas—a modification of the mass wall near outside glass—so that nighttime reradiation from the surface exposed to the sun is directed back into interior spaces. Additional heat storage is provided by a rock bed, which is dampered for two-way flow and charged and discharged by the furnace fan. The duct and flue layout has been carefully arranged to minimize length of runs.

The plan itself is adaptable to various room options, including a bedroom on the lower floor and several entry arrangements. For this particular installation, the lower room will be used as a parlor, and a garage will be added at a later date. The structural framing is carefully rationalized for plywood module construction, and standard lumber dimensions are used throughout. The shape accommodated the hurricane-resistant design required at this site on the Atlantic coast. Its shed roof and shingled exterior are compatible with the traditional house shapes of the area.

Thermal storage is provided by a central masonry wall between living and dining areas. Sunkissed.

From the Inhabitant

In a conventional house, you are living in an atmosphere where temperature is artificially maintained at a particular point most of the time. In a solar house, you are continually aware of, and definitely participating in, the weather cycles of sunlight, temperature, and wind. You become a dynamic part of the interior atmosphere of the house as it changes throughout the day.

When there is bright sun, the first duty of the day is to open the insulating window shades to begin collecting heat. They are left open until the sunlight is gone. If the weather is dark and rainy they are left closed to limit the escape of warm inside air through the windows.

My house warms rapidly on sunny winter days. It goes from a low of 60° F (15.5° C) early in the morning to 70° F (21° C) by 10 A.M., where it remains until nightfall. The rock storage continues to be charged until about 2 P.M. The wood stove is a welcome supplement on cold winter nights. However, this is only necessary during the four coldest months of the year. Otherwise, the manipulation of the insulating window shades seems sufficient to maintain a comfortable temperature during the day, and the furnace comes on only late at night or early in the morning. We use very little auxiliary fuel oil.

The passive solar water-heating system works well. The backup for this system is electrical, and my highest total electric bill in any month has been $24.

Passive solar heat is exceedingly comfortable because it tends to warm people and objects rather than just air. The whole house is light and airy, and no artificial lighting is necessary during the day.

Since the house was built, the basic investment has become more practical because of inflation, and now I anticipate the investment cost of the passive solar system will be paid off in energy savings within a five-year span, rather than the original ten-year payback we had projected.

South-facing windows capture solar energy and coastal view. Sunkissed.

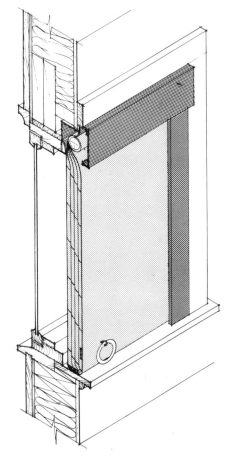

Well-detailed insulative shades are concealed in overhead valence when open. Side track minimizes edge loss.

1. winter solar gain
2. radiant thermal storage mass
3. natural convection
4. forced air to rockbed
5. forced air from rockbed
6. insulated shade
7. shading
8. energy-efficient wood stove

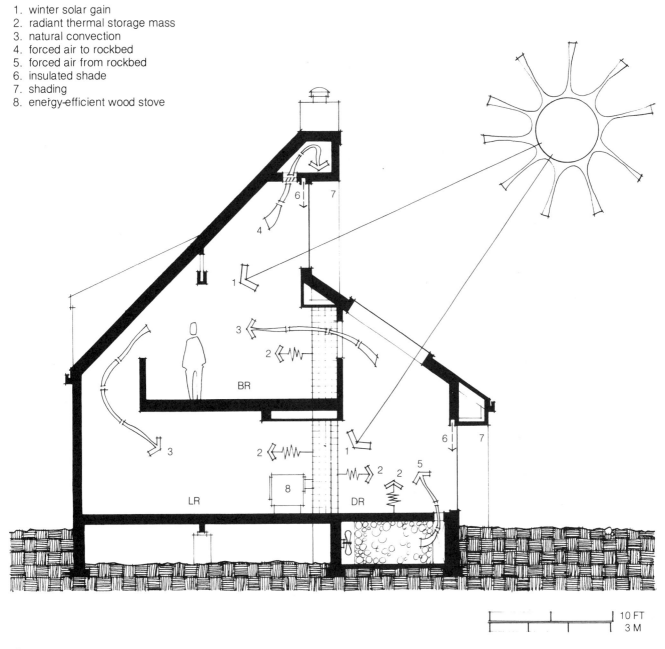

10 FT
3 M

Section

Project Data Summary

Project Information

Project: Sunkissed
    York, Maine
Architect: Donald Watson, AIA
    Guilford, Connecticut
Builder: Mike Fernald
    York, Maine

Climate Data

Latitude	43.5 °N
Elevation	15 FT (4.5m)
Heating degree days	7,446
Cooling degree days	252
Annual percent possible sunshine	59%
January percent possible sunshine	55%
January mean minimum outdoor air temperature	11°F (−11.5°C)
January mean maximum outdoor air temperature	32°F (0°C)
July mean minimum outdoor air temperature	57°F (14°C)
July mean maximum outdoor air temperature	83°F (28°C)

Climate features: severe wind and fog

Building and System Data

Heated floor area	2,000 FT^2 (186m^2)
Solar glazing area	
Direct gain	400 FT^2 (37m^2)
Thermal storage heat capacity	
Tiled masonry chimney	3,150 Btu/°F
Rock bed	19,360 Btu/°F

Performance Data

Building load factor	6.9 Btu/DAY°F FT^2
Auxiliary energy (heating)	45.0 MMBtu/YR
Auxiliary energy (cooling)	1.1 MMBtu/YR
Solar heating fraction	51%
Night ventilation cooling fraction	91%

HODGES RESIDENCE: Ames, Iowa

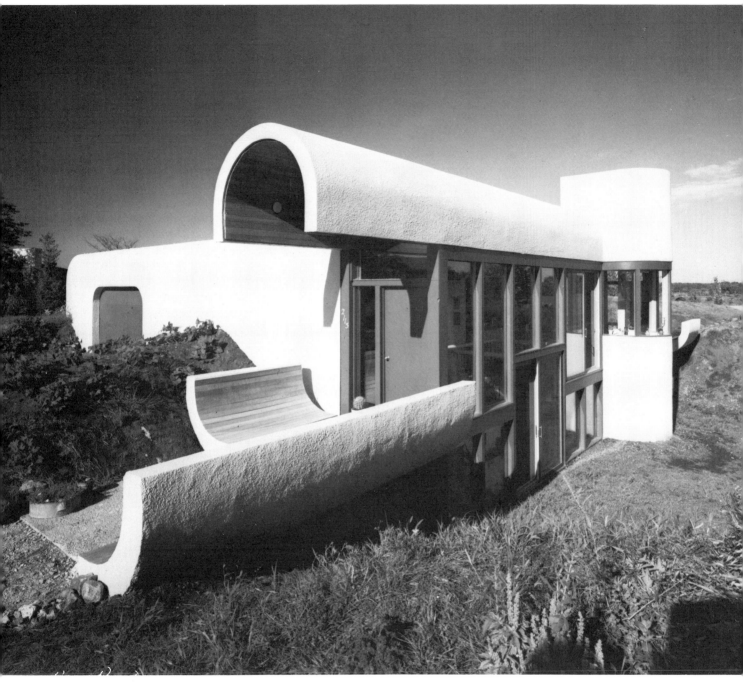

Crisp geometric forms spring from and burrow into the earth. Hodges residence,
Ames, Iowa.

The agriculturally rich Midwest is an ideal proving ground for the principles of passive solar design. The sun's seasonal dynamic is clearly understood in the farmlands, and this ebb and flow of solar power brings rewards to the industrious and observant. Farmers are, for the most part, nature-loving and extremely realistic people. If you can convince a farmer that something is economical and practical, he will see the beauty in it.

Architect David Block designs homes for these farmers and other Iowans who respect nature. His statements are bold and crisp geometric forms that spring from and sometimes burrow into the earth. His architecture develops a unique expression by treating the solar elements as individual features which stand alone yet blend as a part of the overall composition, becoming rational architecture and functional sculpture.

In the Hodges residence, the passive solar system consists of the south glazing, nighttime window insulation (yet to be installed), and concrete thermal storage mass inside the house. The most unique aspect of the design is the cored-concrete slab which forms the floor between the two levels. This slab consists of 8-inch (20cm) thick prefabricated concrete planks with full-length hollow cores. The planks are capped with a 2-inch (5cm) exposed aggregate concrete flooring.

The cores in the slab serve as supply and return ducts for the forced-air system. During the heating season, the small fan on the auxiliary gas furnace circulates air from south to north by day and in the opposite direction by night. During the day, sun-warmed air is pulled through the cores, heating the slab. The slab radiates this heat upward and downward. At night, the fan moves warm air in the opposite direction; when necessary, furnace heat is added to augment the stored solar energy. This interesting use of a horizontal structural slab is an effective and unobtrusive integration of the thermal storage mass. Besides being a basic element in the solar function of the house, the slab also acts as the roof for the tornado shelter of the lower floor.

If one can see the beauty of a well-engineered tool or machine that tills the fields, one can also appreciate the strong, clean lines of a logically designed passive solar home.

Bold curves for the American heartland. Hodges residence.

From the Architect

Passive solar design offers an extremely attractive alternative to architects, builders, and home owners in Iowa. We have a reasonably abundant source of sunlight as well as a substantial heating load for the sun to overcome. Most of the passive solar elements used in the house for heating, cooling, lighting, and hot water operate to near full capacity much of the year. Solar energy is very useful and economical in this region.

Because Iowa has a low population density and building sites tend to be a little larger than in urban areas, we can spread houses out on the land and do more earth berming. The combination of earth integration and passive design is a natural for people here. Many Iowans are agriculturally oriented and close to the land, which makes them appreciate sunlight and landscape.

Passive solar architecture has been greeted with open arms, and the general public is more than willing to invest a small amount extra for the long-term savings. The main problem I have had is justifying the cost of the necessary thermal mass. This design incorporates the mass as a structural part of the house, a twofold investment.

It is my philosophy that passive solar features improve the living environment of my designs. Direct-gain designs allow an open, sunny interior that is both attractive and functional. The solar features, such as glass and thermal mass, are major design elements of the house, used as motifs that add to the honesty of the structure. I feel that the passive solar aspects should be expressed with boldness and should not be covered up or disguised to look conventional.

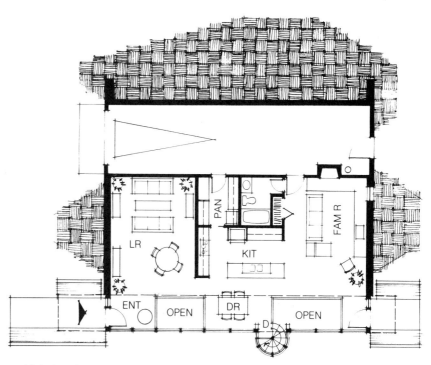

Main level

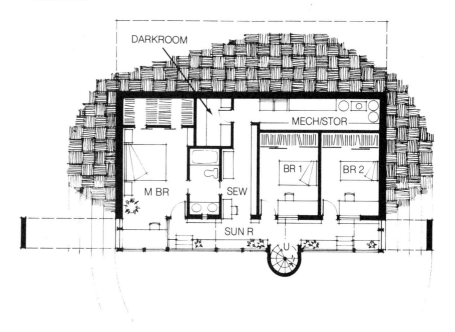

Lower level

3°

10 FT
3 M

From the Inhabitant

Designing, building, and living in a solar home had long been one of our dreams. We felt that nonrenewable fossil fuels were too valuable for other purposes to be wasted in the furnace of an energy-inefficient home, and that Iowans ought to be making better use of solar energy. Why depend on other states and countries for nearly all our energy when we have a vast energy resource of our own? We studied several solar options and rejected active solar systems as too complicated and expensive. Finally, we were convinced that passive solar heating systems could be adapted to our climate.

We decided to build a two-level house with the kitchen, living room, dining room, and family room upstairs.

Cored-concrete slab floor stores solar energy to radiate heat both up and down. Hodges residence.

The bedrooms are downstairs. The lower level remains a few degrees cooler than the upper level. Bedrooms were built downstairs because they could remain cooler in winter, should be cooler in summer, and were best downstairs for tornado protection.

The south wall is almost all glass to admit large amounts of solar heat during the winter, when the sun stays low in the southern sky. To keep the house from overheating on sunny days, a concrete floor was built between the two levels. It absorbs solar heat during the day and releases it at night when the house starts to cool down. This helps maintain a stable inside temperature. Windows on the north, east, and west sides of the house were avoided, since they collect very little solar heat in winter and lots of it in summer, when it is not wanted.

The house was very comfortable during the winter, with the temperature 65°F and warmer. This is because of the radiant heat from the warm floor and walls of the well-insulated house. On sunny winter days the temperature often reaches the high 70s and allows sunbathing by the windows, a wonderful luxury that makes us look forward to those bitterly cold but sunny midwinter days.

I had calculated that our total heating bill for the gas backup furnace would be only $20 to $40 for a whole winter if we insulated our windows at night—if not, it would be $80 to $120. The first winter we did not use window insulation and we used $85 worth of gas for heating, as measured by the furnace meter. We also burned about 1,200 pounds of scrap wood in our fireplace, saving us about $10 worth of gas. We were very pleased with a total annual heating bill under $100, even though this first winter was one of the cloudiest on record.

In winter, we greatly enjoy our exposed aggregate concrete floor, which has a pleasing appearance. It remains warm during the winter and is fun to walk barefoot on. In summer, the upper level becomes uncomfortable after several days of hot and humid weather. We intend to improve conditions by increasing natural ventilation, using louvers and screen doors, and adding exterior summer overhangs on the south side. The lower level has remained cool even in hot periods, but fans are sometimes necessary to combat humidity.

These simple passive principles can be used in any home, whether large or small, plain or fancy. We now know from direct experience that our solar home design is one possible approach that works well in Iowa.

1. winter solar gain
2. warm air through slab to furnace
3. fan unit
4. air return to space
5. radiant thermal storage mass
6. insulated shade
7. north buffer zone
8. earth integration
9. energy-efficient wood stove
10. ventilation

Open sunny interior is both practical and beautiful. Hodges residence.

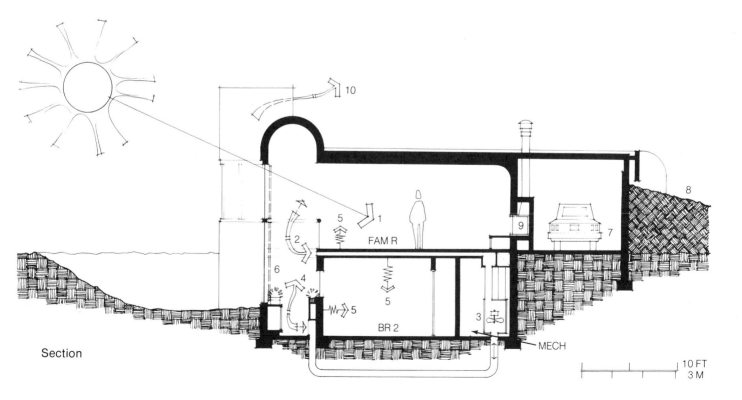

Section

10 FT
3 M

Project Data Summary

Project Information

Project: Hodges residence
        Ames, Iowa
Architect: David Block, AIA
        Ames, Iowa
Builder: Laurent and Linda Hodges
        Ames, Iowa

Climate Data

Latitude	42.0°N
Elevation	975 FT (297m)
Heating degree days	6,800
Cooling degree days	948
Annual percent possible sunshine	61%
January percent possible sunshine	53%
January mean minimum outdoor air temperature	11°F (−11.5°C)
January mean maximum outdoor air temperature	30°F (−1°C)
July mean minimum outdoor air temperature	63°F (17°C)
July mean maximum outdoor air temperature	88°F (31°C)

Climate features: cold, dry winters; hot, wet summers with tornadoes

Building and System Data

Heated floor area	2,200 FT^2 (204m^2)
Solar glazing area	
Direct gain	494 FT^2 (46m^2)
Thermal storage heat capacity	
Cored concrete floor	18,480 Btu/°F

Performance Data

Building load factor	7.0 Btu/DAY°F FT^2
Auxiliary energy (heating)	34.4 MMBtu/YR
Auxiliary energy (cooling)	1.2 MMBtu/YR
Solar heating fraction	76%
Night ventilation cooling fraction	92%

ABERCROMBIE RESIDENCE: Salina, Kansas

Berming and shading louvers for solar housing in middle America. Abercrombie residence, Salina, Kansas.

Much of middle America has found that solar makes sense. This Kansas residence, designed and built by C. F. Abercrombie, responds well to the strong forces of the climate. Hot summers, cold winters, and tornadoes are common, and this earth-sheltered home is designed to take it through all seasons.

The house makes efficient use of space to allow a comfortable living and dining area, along with two bedrooms and a loft study. Each major space, with the exception of the back bedroom, opens visually and functionally to the south patio. The fifty-degree sloped glass is intended to increase winter gain into these spaces. Typi-cally, sloping glass creates a problem, with both sealing and shading. Abercrombie has provided fixed louvers to provide summer shading while allowing winter light to enter high into the space. All glazing is provided with automatically pivoting interior insulating louvers to minimize winter heat loss.

An important feature of the house is its siting. It is well bermed in the north, east, and west, affording minimum exposure to temperature and winds. Located in a middle-class neighborhood, the house was intentionally designed for marketability, with style and function that cater to Midwest tastes. It is a structure that compares in size and features to its neighbors. Except for unusual exterior elements such as shading louvers and earth berming, it uses materials and forms similar to other nearby homes. Yet the energy required to heat and cool this house is substantially reduced.

The design of this house deals directly with the environment, bringing a new kind of logic and beauty to the heart of our country.

From the Designer

The basic design concept for this home was to optimize passive solar design principles. The plan's north-south to east-west dimensions have a ratio of 1.5:1, and it is oriented fifteen degrees east of true south (following Victor Olgyay's recommendations in *Design with Climate*). Heat-producing spaces such as the kitchen, machine room, laundry, and bathrooms are located in areas receiving reduced sunlight helping to balance internal and external gain. An entry vestibule for infiltration control and a garage on the north act as thermal buffer zones. Penetrations are minimized on north, east, and west wall surfaces, reducing exposure to adverse temperature, wind, and solar conditions.

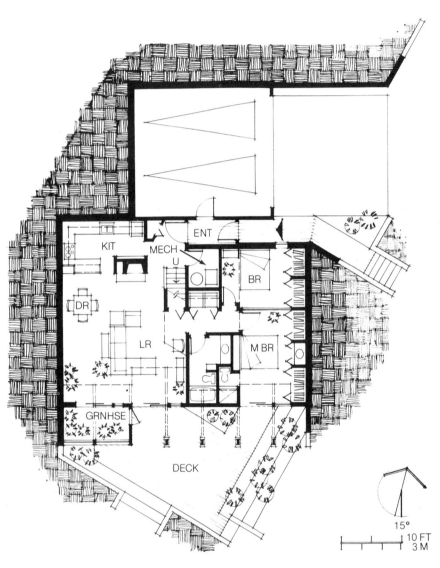

Floor plan

15°

10 FT
3 M

Attention was given to construction details with the use of the following:

1. External shade louvers
2. Thermal breaks between highly conductive materials and their exposure to the outside
3. Movable insulation on all glazed openings
4. Insulating glass throughout
5. Wood-framed windows and a thermal-break (low heat conductivity) skylighting system

Problems commonly associated with passive design, such as overheating and glare, are also dealt with. The glazing was reduced to the optimum by equating the solar collection area with estimated heating loads. Overheating is reduced by the following:

1. Generous use of concrete thermal storage mass throughout the house
2. Shading of the fixed exterior louvers
3. Adjusting the pivoting interior louvers, which also control glare
4. Closing the window insulation when needed to prevent heat gain
5. Opening the greenhouse to the outside to dump heat
6. Using the ceiling fan to circulate inside air

This design illustrates that the requirements imposed by the use of passive design tools can also improve the esthetic conditions of a living environment. The ability to upgrade the spatial quality of residential housing, and at the same time drastically reduce energy demands, is an exciting and worthwhile endeavor.

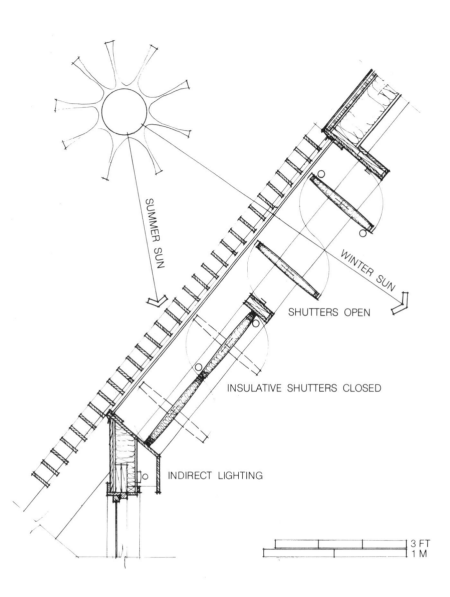

Detail of insulated louver and summer sunshade

Insulating interior skylights minimize winter heat loss. Abercrombie residence.

From the Inhabitant

This home is a statement of determination to conserve energy and integrate with the surrounding environment. It provides an atmosphere of simplicity, openness, and efficiency. Living in a passive solar home requires getting involved directly with the living environment. Opening and shutting movable insulation, adding or removing clothing, and other participation in the operation make you a part of the house instead of imposing yourself on it.

We hope to set an example for other people by demonstrating that homes like ours, for people of average means, can be beautiful and sensible. We hope to spread the word, in our own way, that we can all live in an energy-efficient and environmentally conscious world. If we learn to appreciate our environment by a more involved relationship with it, we will also learn to conserve our natural energy sources.

1. winter solar gain
2. radiant thermal storage mass
3. shading
4. insulated louvers
5. insulated shutters
6. greenhouse gain
7. ceiling fan
8. earth integration

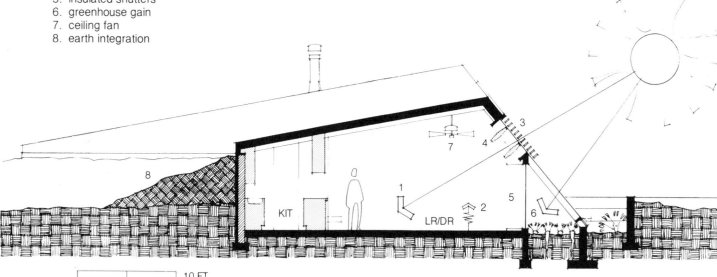

10 FT
3 M

Section

Project Data Summary

Project Information

Project: Abercrombie residence
Salina, Kansas
Designer/builder: C. Fred Abercrombie
Salina, Kansas

Climate Data

Latitude	38.8 °N
Elevation	1,272FT (367.5m)
Heating degree days	5,628
Cooling degree days	1,302
Annual percent possible sunshine	67%
January percent possible sunshine	60%
January mean minimum outdoor air temperature	20°F (−6.5°C)
January mean maximum outdoor air temperature	42°F (5.5°C)
July mean minimum outdoor air temperature	68°F (20°C)
July mean maximum outdoor air temperature	95°F (35°C)

Climate features: moderately cold winters; mild summers

Building and System Data

Heated floor area	1,412 FT^2 (131m^2)
Solar glazing area	
Direct gain	364 FT^2 (34m^2)
Greenhouse	109 FT^2 (10m^2)
Thermal storage heat capacity	
Tiled concrete floor	20,296 Btu/°F
Concrete walls	5,801 Btu/°F

Solar clerestories repeat the jagged form of the mountains. Stockebrand residence, Albuquerque, New Mexico.

Mazria/Schiff & Associates has followed the ideal of refining the design vocabulary of passive solar for some time. This New Mexico residence combines passive solar principles and contemporary southwestern styling into a well-balanced architectural statement.

The design differs from the stereotyped idea of the south-facing, direct-gain glass wall. Each room of the house steps around the southwest side of a swimming pool to take advantage of the view. South- and southeast-facing clerestory monitors reach elegantly up to capture solar gain. This juxtaposition of elements creates a dynamic form which responds to functional, esthetic, and thermal needs.

The swimming pool is a vital part of the interior, fulfilling several needs aside from recreational. Each living space is thermally linked to this plant-filled oasis. It satisfies a desire for moisture and vegetation in this dry, high desert climate. The pool also assists with space conditioning by capturing and holding heat and adding humidity to the entire house.

Integration of thermal with other architectural considerations is the design challenge in passive solar housing. The Stockebrand residence exhibits the organizational elements that are part of the planning and proportioning aspects of passive design. Solar glazing, interior thermal mass, and exterior insulation are three variables of any given space. The architects have gone beyond simply sizing these basics to the more complex organization of the individual volumes into an architectural whole. Each space is located within the total volume to best suit its use, both thermally and functionally.

South- and southeast-facing clerestories reach up for solar gain. Stockebrand residence.

Contemporary southwestern design incorporates a swimming pool for thermal storage. Stockebrand residence.

From the Architects

The client, a mechanical engineer, came to us with the desire for a house that would rely minimally on the local public utility by being totally solar heated and cooled. The functional program was simple enough: a one-level house for a family of five with visually connected living spaces, acoustical and physical privacy between the adult and children's areas, and a well-integrated enclosed swimming pool. The site, in the foothills of the Sandia Mountain range east of Albuquerque, New Mexico, slopes to the southwest and is exposed and barren except for small, hardy cedar and sage trees between granite outcroppings.

We felt strongly that a house in this exposed terrain should hug the ground, both to intrude as little as possible on the soft profile of the foothills and to protect the house from strong winter winds. To that end, the building was designed with flat roofs and bermed into the southwest slope so that the roof line as seen from the road and neighboring houses to the north is low to the terrain. The south face of the building is staggered to define and protect outdoor places, which are covered by trellises. By locating connections to the site through these outdoor "rooms," a gradual transition is achieved between the sheltered interior and the rather austere, awesome expanse of desert outside. The clerestories projecting from the low roof repeat the jagged form of the mountain peaks behind them.

Circulation through the house occurs along the edge between the living spaces and the pool enclosure. The living spaces, grouped along the south face of the building at a forty-five-degree angle to the pool, admit direct gain in winter and boast views of the city 1,000 feet (305m) below to the southwest. The pool enclosure buffers the living spaces from the cold north exposure. It is heated primarily by direct gain through large south-facing sawtooth clerestories, and to some extent by the pool water, which will be heated to 83°F (28°C) by a flat-plate solar heating system. Since the pool is constantly losing heat to the pool enclosure, and this heat is transferred through an uninsulated wall between the pool area and the house, the pool's solar heating system will serve as a backup heating system for the entire house.

Sunlight enters each living space in two ways: through low-view windows to the southeast and southwest and through clerestory windows facing east of south. This allows for balanced daylight that reduces glare, as well as providing for winter heat gain. Sunlight reflecting from light-colored interior masonry walls is distributed throughout each space for effective absorption and storage by the masonry walls and brick floors. The exterior walls are constructed of concrete block, insulated on the outside and then stuccoed. In effect, each space becomes a live-in solar collector, storage, and distribution system. The large masonry weatherskin in this expansive one-story house is possible in this sunny, temperate climate without penalty of excessive heat loss, since most of the exposed skin faces south.

Summer cooling is accomplished by the same elements as winter heating: masonry thermal mass and glazing. Masonry, distributed throughout the house, is cooled by ventilating the building with cool night air. All glass areas are shaded with overhangs or vegetation to minimize daytime solar gain in summer.

The image of this passive solar house is one of clusters of forms protecting and opening onto outdoor spaces, not unlike the puebloes found in this region.

From the Inhabitant (after three months)

Bright, light, airy—that describes the feeling of the house. The clerestory windows and the exposed beams open up the house and give it a feeling of space. The large amount of glazing, had it been at usual window heights, would have burned us out on bright winter days.

Open, yet private. This paradox occurs because, while the house has an open floor plan, the masonry walls absorb and diffuse sound so that it is not heard more than one room away. All parts of the house—with the exception of the master bedroom and lofts, which get cool—have fairly even temperatures, with few drafts or cold spots.

In general, the house is just great to live in. The kids say, "warm, fun, neat, spacious, light, bright." The views are terrific. The architects did an excellent job on the area lighting—both the cove lights and the fixtures are dramatic.

Our lives haven't changed much, though we are more aware of the

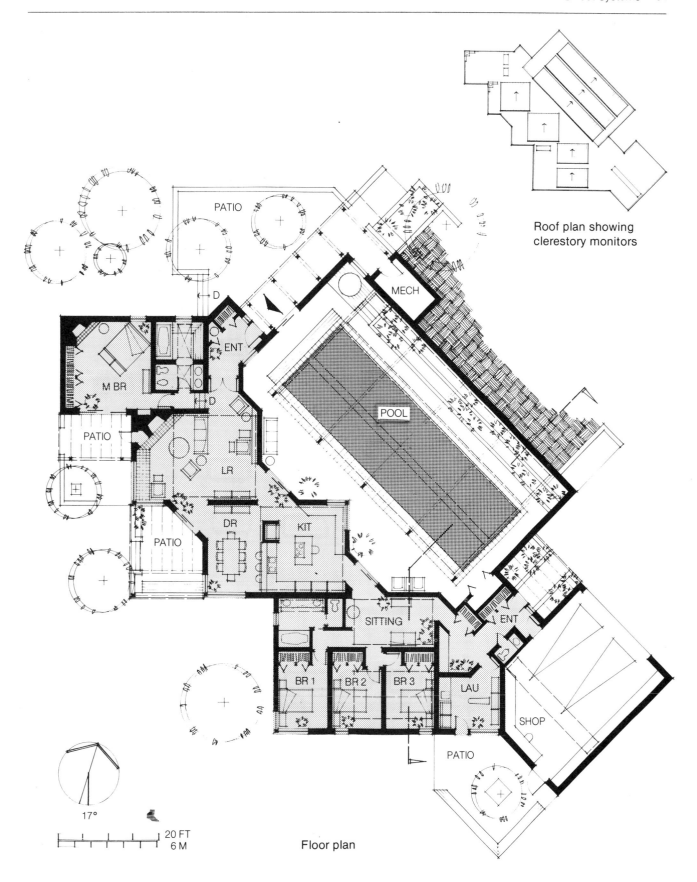

Roof plan showing
clerestory monitors

PATIO

MECH

ENT

M BR

D

PATIO

LR

POOL

PATIO

DR

KIT

SITTING

ENT

BR 1

BR 2

BR 3

LAU

SHOP

PATIO

17°

20 FT
6 M

Floor plan

changes of seasons and the weather in general. In winter, we tend to wear less clothing because of the evenness of temperature and the privacy. Conversely, because of the overhangs, the house is actually a little too cool on slightly overcast days in early summer. Few nights are cold enough to require sweaters, but some early mornings feel slightly cool for the season.

Most of the bedrooms approximate conditions in the rest of the house, but the master bedroom runs too cool in Feburary, March, and April. This may need extra attention. We may also need more window covering after the sun goes down, but we want to keep the view. Also, stratification of warm air to the lofts is raising havoc with people's ability to sleep late in the mornings. The sewing room, which is cold, should have had the clerestory extended over it.

Once we install the active solar pool backup system, we should be able to reduce our electrical consumption to just lights and appliances. At that point, it will be time to try solar cells to finish the job of being electrically independent.

Solar gain through windows and high clerestories allows balanced daylight and winter heating. Stockebrand residence.

1. winter solar gain
2. radiant thermal storage mass
3. earth integration
4. clerestory solar monitors

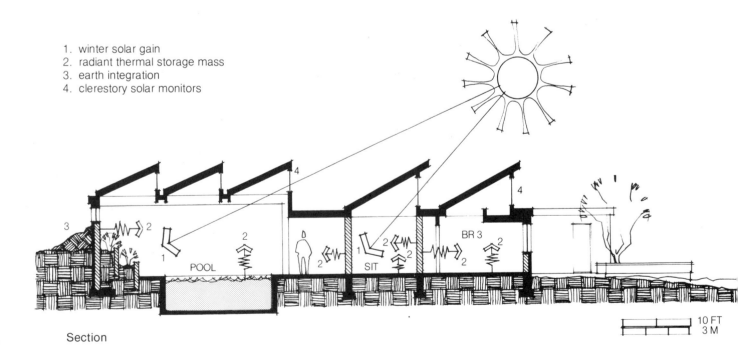

Section

Project Data Summary

Project Information

Project: Stockebrand residence
          Albuquerque, New Mexico
Architect: Mazria/Schiff, AIA & Associates
          Albuquerque, New Mexico
Designers: Edward Mazria, Mark Schiff, Rob Strell
          Albuquerque, New Mexico
Builder: Jim Bishop Construction
          Albuquerque, New Mexico

Climate Data

Latitude	35.0°N
Elevation	6,280 FT (1,914m)
Heating degree days	5,500
Cooling degree days	1,000
Annual percent possible sunshine	77%
January percent possible sunshine	73%
January mean minimum outdoor air temperature	23°F (5°C)
January mean maximum outdoor air temperature	46°F (7.5°C)
July mean minimum outdoor air temperature	65°F (18.5°C)
July mean maximum outdoor air temperature	91°F (32.5°C)

Climate features: cold, dry, windy winters; hot, dry summers

Building and System Data

Heated floor area (including pool)	4,764 FT² (442.5m²)
Solar glazing area	
Living spaces	595 FT² (55m²)
Pool area	720 FT² (67m²)
Thermal storage heat capacity	
Brick on sand and concrete tiled floor	47,160 Btu/°F
Interior masonry walls	77,190 Btu/°F

This small solar box works well in the cloudy Pacific Northwest. Hart residence, Portland, Oregon.

Lower costs are difficult but possible through simplicity. Hart residence.

The Pacific Northwest is not well known for its abundant sunshine; in fact, many people associate this area with cloudy days and drizzle. But architect William Church of Portland, Oregon, has a different vision. He can't stop the rain, but he has designed passive solar houses that work there.

The cloud cover and resulting diffuse solar energy are normally accompanied by moderate temperatures, the clouds acting as an insulative blanket which reduces surface radiation from the earth to space. When the sky is overcast, the average winter temperatures are normally in the 40–50°F (4.5–10°C) range. Naturally, the amount of incoming annual solar radiation is reduced, but so is the inside-to-outside temperature differential, resulting in a lower building heat loss. Accordingly, passive solar design is just as feasible here as in colder climates with more solar radiation: less heating often requires less solar gain.

This economical little box is a studied exercise in utilizing standard materials, ordinary construction techniques, and basic solar design principles. It achieves maximum results for minimum cost. Construction budget is a constant battle for most architects. This austere design is a good example of productive collaboration

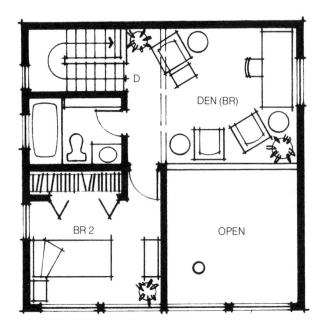

Upper level

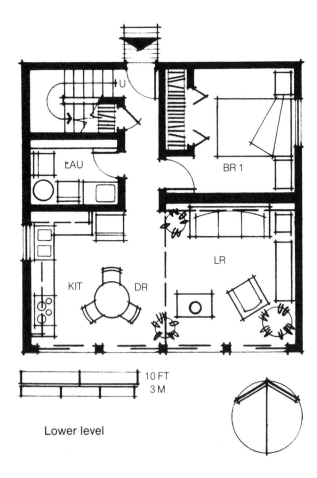

10 FT
3 M

Lower level

between owner, architect, and contractor. Each is delighted at having been able to come up with a custom-built passive solar house which cost less to build than a standard tract house. The direct-gain south glazing, interior mass floors and wall, and wood-burning stove are all that are needed to make this inexpensive, contemporary home a thermally comfortable place to live.

From the Architect

To put it simply, this house was to cost as little as possible, have passive solar heating, and be livable. It couldn't have happened without the willingness of the Harts to make it work and the gutsiness of the contractor to give it a try. I feel very good about the results; the things learned far outweighed the usual hassles of building a house (easy to say now!). What was learned?

1. Lower costs are difficult but possible—through simplicity.
2. Passive features can be accomplished with little or no increase in costs—through careful planning.
3. This house has helped further my thinking about the balance between conservation and solar design in this climate—both work.

In short, I love this little solar box.

From the Inhabitant

While living on a boat docked at a river island near Portland, we became aware of the first addition to our family. A sailboat is a fine adventure home for a couple, but certainly not a suitable place to expand for a new baby. The cost of buying a decent home was financially beyond what we were willing to invest. We decided to buy a piece of land and build our own house for a combined total of $40,000. With luck, we were able to buy some wooded acreage near Portland for $15,000. That left $25,000 to build with.

We were interested in a solar home for both economic independence and energy conservation. The local utility company had built an active solar demonstration home here a couple of years ago, the basement of which looks like the engine room of a ship. That was not for us. While researching solar-home feasibility in the area, we were able to obtain a list of local architects with experience in designing solar buildings. Passive solar seemed to offer the most in terms of economy, simplicity, and efficiency.

We selected Bill Church because he expressed enthusiasm for the challenge. Also, he had helped design a small cooperative community which consisted of a number of economical passive solar homes. These homes had cost slightly more than we were planning to spend, but they were built on a sloping site. Bill figured that with some design simplification and by involving a contractor early on, we could reduce the construction cost to rock bottom.

Living on the sailboat had taught us a great deal about efficient use of space. We settled on a two-bedroom, one-bath plan of about 1,000 square feet (93m²) of floor area. With frugality in mind, we planned the design around a number of economy measures: the

Brick-on-concrete floor and masonry collect and store the sun's heat. Hart residence.

site was flat, two-story construction reduced foundation cost, a flat roof was cheaper than a pitched one, window units were selected for maximum glass area at minimum cost, structural spans were suited to normal lumber lengths, and all materials were standard modular units. The items that were not compromised were 6-inch (15cm) thick frame wall construction for better thermal insulation and increased passive solar performance. The house was sited facing due south, and the trees were cleared away to allow low winter sun to strike the south glazing.

The initial construction bid was $38,000. After working with the contractor, we took on several tasks which lowered the cost to $28,000. We agreed to clear the site, install the temporary power pole, buy a few appliances, lay the brick floor, and paint, among other things. The contractor was very helpful. As a team, we and the architect achieved a final cost of $32,000, including all fees.

The project has been a success. An unexpected bonus is the quality of natural lighting inside, wonderful in this cloudy climate. We like the open spaces after living on a boat. The lack of sound privacy between rooms is something we have learned to live with. The thermal mass consists of the brick-on-concrete slab floor and interior concrete block wall. Some day, we will plaster the wall. The sixteen-foot-high ceiling in a relatively small living space is a bit too high. Lowering the ceiling fan and light fixtures should improve the room's appearance.

Thermally, the house performs well. Actually, it is cooler in the summer than we expected. Cooling is accomplished by opening windows, closing blinds, and shading by the outside trees. In winter, the wood stove is our only auxiliary heat source. The reflective metal blinds are closed at night to somewhat reduce heat loss through the glass. The house is easy to heat and the mass wall retains its warmth well. Each space has a different temperature characteristic. The back or north rooms are harder to heat, and the upstairs is generally warmer than downstairs. We don't use the ceiling fan, which blows warm stratified air down; it works, but we're generally comfortable without it. Last winter, we burned slightly more than a cord of maple. This year, we will burn oak and expect to use even less.

1. winter solar gain
2. radiant thermal storage mass
3. sun-control blinds
4. deciduous trees
5. energy-efficient wood stove

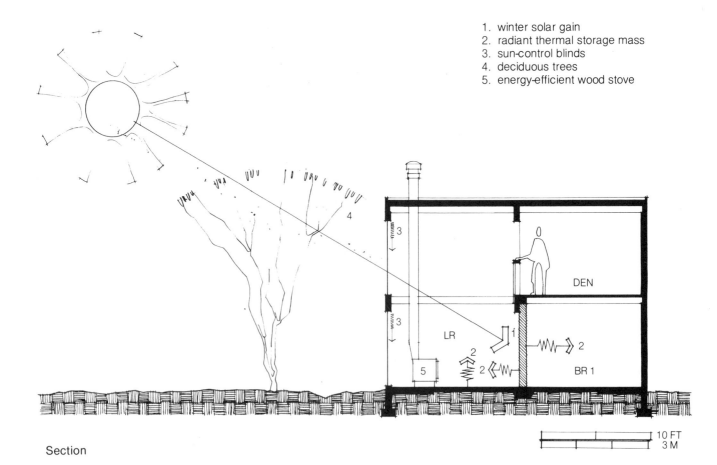

Section

10 FT
3 M

Project Data Summary

Project Information

Project: Hart residence
            Portland, Oregon
Architect: William Church, AIA
             Portland, Oregon
Builder: Prinz Masterson
           Portland, Oregon

Climate Data

Latitude	45.6°N
Elevation	100 FT (30.5m)
Heating degree days	4,700
Cooling degree days	300
Annual percent possible sunshine	44%
January percent possible sunshine	21%
January mean minimum outdoor air temperature	33°F (.5°C)
January mean maximum outdoor air temperature	44°F (6.5°C)
July mean minimum outdoor air temperature	55°F (12.5°C)
July mean maximum outdoor air temperature	79°F (26.1°C)

Climate features: overcast winters; mild summers

Building and System Data

Heated floor area	954 FT² (88.5m²)
Solar glazing area	
Direct gain	267 FT² (25m²)
Thermal storage heat capacity	
Brick-on-concrete floor	6,610 Btu/°F
Interior masonry wall	1,840 Btu/°F

"Ears," louvers, and drums are this house's striking solar features. Parham residence, Randleman, North Carolina.

Fifty-three acres near Raleigh, North Carolina, have been set aside by a small group of people as an energy-conscious community. Up to six solar residences will be built for this demonstration project, which has allocated 40 percent of the land for private ownership and set the remainder aside as common open space. This is a very picturesque setting for energy-conscious rural living.

One house already completed is the Parham residence. It was originally designed on speculation for resale. The designers, John Alt and Scarlett Breeding, built what looks to become a showcase of passive design for this community. By taking several specific functional issues and

North side shows variety and textural relief. Parham residence.

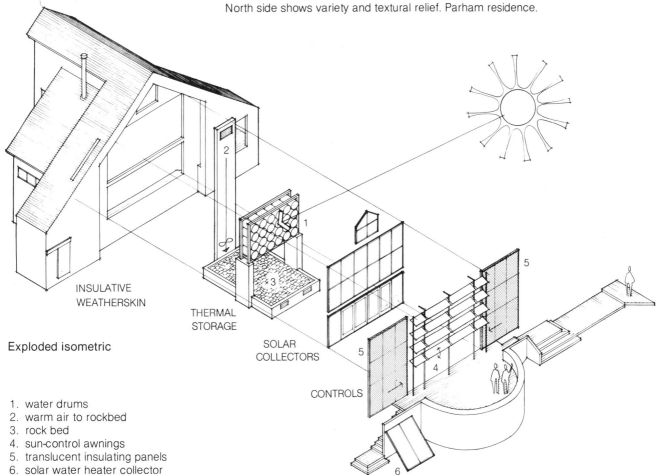

INSULATIVE WEATHERSKIN

THERMAL STORAGE

SOLAR COLLECTORS

CONTROLS

Exploded isometric

1. water drums
2. warm air to rockbed
3. rock bed
4. sun-control awnings
5. translucent insulating panels
6. solar water heater collector

dealing with them directly, Alt and Breeding have glamorized classical water-drum thermal storage, flexible shading, and movable insulation. Each of these passive solar devices has been handled with strong yet delicate detailing to the point that they have become vivid architectural expressions.

The house is essentially a direct gain–water wall concept, employing twenty-four fifty-five-gallon custom-fabricated drums. These are mounted in a custom frame and set away from the south glass toward the building center, becoming a major sculpture in the space. Since the water's heat-carrying capacity is slightly less than the total gain on a winter day, a rock bed was built below the wood floor for additional thermal storage. Stratified warm air is transferred by fan from the third level to this secondary storage only when the temperature of the upper space exceeds the comfort level.

All south windows are single-glazed for maximum solar gain. To overcome excessive losses through this wall of glass, a translucent insulating panel rolls to cover its exterior. When open, the panels appear from a distance as giant ears above the gable roof. The translucency of the double-skinned fiber-glass panels offers a tight weatherskin while allowing light to enter the space.

One of the most striking features of the house is the sun-control awnings over the south glass. Two-story glaz-ing is difficult to shade at any time during the year; often, fixed overhangs will not let in enough sunlight during cooler spring days, or will allow too much gain in early fall, when cooling is still a problem. This house displays a solution to each of these problems. Like a human eyelid, a series of awnings is controllable and adjustable from both inside and out, allowing the user total control of sunlight while maintaining the view.

The visual and functional strength of these elements is a vital force in

EXTERIOR CONTROL

INTERIOR CONTROL

Sun-control awnings are controlled both inside and out.

Water drums and frame integrate to become a single visual and structural element. Parham residence.

the passive solar vocabulary, just as doors, windows, and fascias are in conventional design. In this design, they make visual sense—and, moreover, they work!

From the Designer

Basically, the design was an exploration of the following issues:

1. How can fifty-five-gallon steel drums, used to store water as thermal mass, be integrated with the interior architectural composition? This is a house built for resale, and the concern was that the drums would be objectionable to most people. Clearly, the drums had to transcend their "drumness," to become something other than what they really are. This was accomplished by treating the drums and their frame as a structural and visual element in the architectural design.

2. How can movable insulation over a very large window wall be made easy and fast to close— an incentive to actually use it? And when it is closed during cloudy weather, is it possible to prevent the house from turning into a dark cave? Lightweight, translucent movable insulating panels answered these questions.

3. With a very high south window wall, how can light penetration and heat gain be controlled at all times of the year, especially during fall and spring, when the sun is low enough for significant heat gain, yet the weather may be quite warm? (During October, we frequently alternate between a few days of 40–50°F [4.5–10°C] weather, then suddenly a few days of 70–80°F [21–26.5°C] weather.) We used easily adjustable horizontal sun-control awnings.

These were some of the basic considerations which generated the design. Since then I have not pursued "solar" in any aggressive way— partly because teaching has left little time for design work, partly because I was beginning to feel pigeonholed as a designer. I find solar to be one of the most interesting aspects of design today—certainly one of the most critical—but not the only issue. For instance, the whole time I was building this house, what consistently went through my mind was not, "Wow, this is going to be a terrific house." Instead, I was becoming more and more appalled at the way we have to build houses today: there's something archaic about it. I sense very strongly that somewhere in the future people are laughing at us. Houses come in about three million parts, none of which fits very well with the other and few of which are locally available. When the house is finished, there is often a pile of waste material as big as the house itself, and with which—if it only weren't in such small pieces—

you could probably construct another house. The complex process of getting a house built is enormously stressful due to the necessary management of many parts, tasks, and unnecessary expenses. These, of course, are not new insights. What is new is that they began to profoundly affect me personally: I cannot face the prospect of building another house, no matter how logical or beautiful or solar, until the *process of designing and building* that house is just as beautiful.

Another reason I no longer stress solar in my design work is that I feel building energy conservation will eventually emerge more as a technological issue than an architectural one. I predict a split in the design and construction processes in which highly engineered, self-sufficient non-specific structures will take over the basic functions of climate control, while "architecture" will become a kind of super interior design. And this is the direction in which I am currently working.

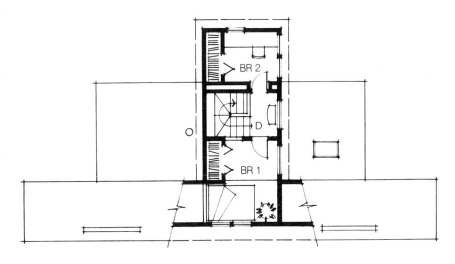

Third level

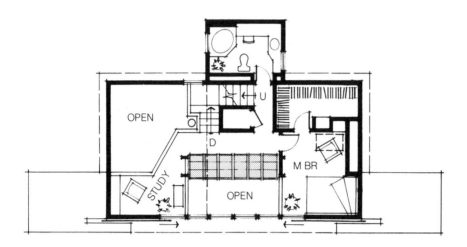

Second level

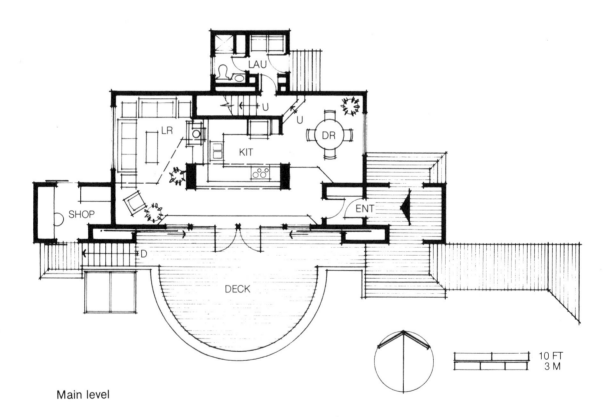

Main level

10 FT
3 M

Project Data Summary

Project Information

Project: Parham residence
          Randleman, North Carolina
Designers: John Alt with Scarlett Breeding
Builder: John Alt

Climate Data

Latitude	36.0 °N
Elevation	434 FT (132m)
Heating degree days	3,810
Cooling degree days	1,349
Annual percent possible sunshine	61%
January percent possible sunshine	56%
January mean minimum outdoor air temperature	32°F (0°C)
January mean maximum outdoor air temperature	53°F (11.5°C)
July mean minimum outdoor air temperature	67°F (19.5°C)
July mean maximum outdoor air temperature	89°F (31.5°C)

Climate features: long, pleasant spring and fall; hot, humid summers; cold, rainy winters

Building and System Data

Heated floor area	1,250 FT^2 (116m^2)
Solar glazing area	
Direct gain	380 FT^2 (35.5m^2)
Thermal storage heat capacity	
Water drums	10,980 Btu/°F
Rock bed	8,000 Btu/°F

Performance Data

Building load factor	7.9 Btu/DAY°F FT^2
Auxiliary energy (heating)	8.3 MMBtu/YR
Auxiliary energy (cooling)	1.2 MMBtu/YR
Solar heating fraction	67%
Night ventilation cooling fraction	84%

Long and low on a dairy pasture. Walker residence, Springfield, Missouri.

The basic principles of solar energy often come second to architectural style. A pleasing exception is the solution which combines both image and thermal function into a graceful whole. Architect James Lambeth has managed to accomplish this in his design for the Walker residence. Its walls and roof are extremely well insulated thermally for this climate. Nearly all of the south wall is glass, and the northeast and west walls are extensively earth-bermed. The plan has placed all the storage and utility spaces in the northern portion of the house as buffer zones and all the living spaces in the southern portion for sun, view, and ventilation.

The house, especially from the north, is long, long, long. Complementing its length is its low profile, which flows with the horizontal, rural countryside. The form is boldly penetrated at the center by an all-glass greenhouse to create a striking entry. This is reminiscent of the traditional "dog run," a narrow hallway at the center of the house, common in the southern states. To create movement and an invitation to the entry, curved sloping walls extend out from the rectangular plane of the house.

From the Architect

The Walker residence is located on a ranch in southern Missouri. The site is a vast pasture used for grazing both horses and cattle. I wanted the home to be a quiet part of this rural environment. The design, which evolved as a long, low profile on the north entry side, is a result of both esthetic and environmental concerns. The low silhouette allows the home to blend into the horizontal plane of the site. Earth-berming extends the visual line of the field to the roof overhang, also providing protection from severe winter winds. The north roof overhang is curved downward to deflect the winter winds. Windows located below this overhang allow ventilation in spring and fall and provide natural lighting to the back interior zone.

All major spaces of the home face south to views across the fields. These double-glazed walls facilitate view and transmit solar gain during the winter. Overhangs protect the glass from the high summer sun.

Black slate covers the floor slab and massive fireplaces to provide

All major spaces orient south to views across the fields. Walker residence.

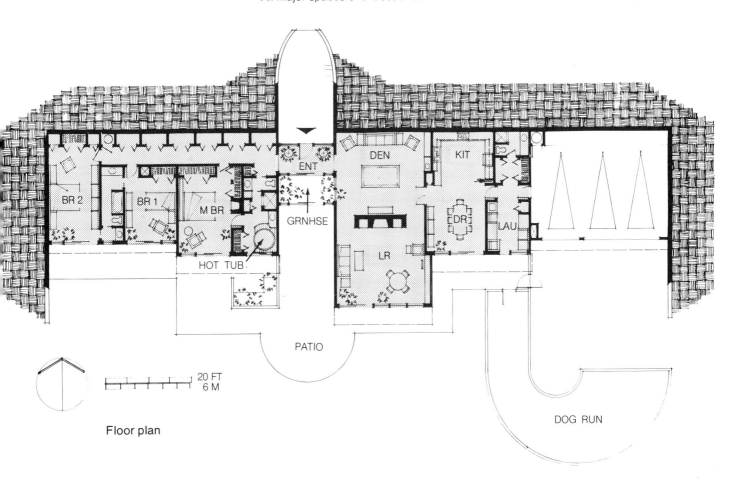

Floor plan

20 FT
6 M

BR 2   BR 1   M BR   ENT   GRNHSE   DEN   KIT   LR   DR   LAU

HOT TUB

PATIO

DOG RUN

Slate-covered fireplace stores heat and provides backup for passive solar heating. Walker residence.

heat storage. The floor slab is insulated from the ground to improve heat retention. The fireplaces and forced-air heat pumps provide backup heating and cooling when needed. Ceiling fans mix the interior air, providing comfort in all seasons. Solar collectors, located at each end of the house, provide energy for two water-heating systems.

My design process involves the interaction of site, client needs, and a kind of personal search for "correctness" in architecture. Energy efficiency is only one of the many varied determinants in the design process. Each site, client, and indigenous material personalizes a particular design to its region. Whether the project is located in Aspen or Las Vegas, the process is the same, but the resulting architecture is totally different. This is what I find constantly fascinating about solar architecture.

1. winter solar gain
2. radiant thermal storage area
3. greenhouse entry beyond
4. shading
5. north roof overhang
6. earth integration
7. vent window

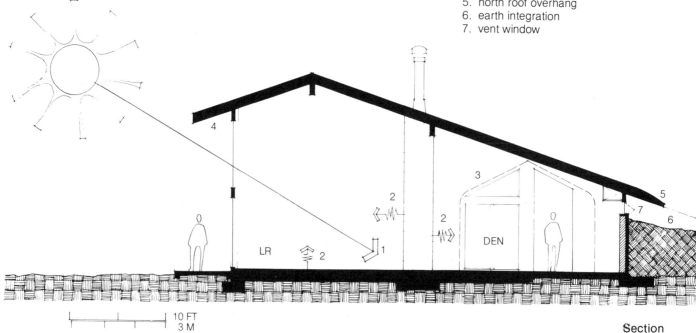

10 FT
3 M

Section

Project Data Summary

Project Information

Project: Walker residence
                Springfield, Missouri
Architect: James Lambeth, AIA
                Fayetteville, Arkansas
Builder: John Coombs
                Springfield, Missouri

Climate Data

Latitude	37.2°N
Elevation	520 FT (158.5m)
Heating degree days	3,800
Cooling degree days	2,420
Annual percent possible sunshine	63%
January percent possible sunshine	50%
January mean minimum outdoor air temperature	24°F (−4.5°C)
January mean maximum outdoor air temperature	43°F (6°C)
July mean minimum outdoor air temperature	67°F (19.5°C)
July mean maximum outdoor air temperature	90°F (32°C)
Climate features: temperate	

Building and System Data

Heated floor area	3,562 FT^2 (331m^2)
Solar glazing area	
Direct gain	902 FT^2 (83.5m^2)
Thermal storage heat capacity	
Slate-covered concrete floor	73,950 Btu/°F
Slate-covered concrete fireplace	13,800 Btu/°F

High temperatures and humidity are the challenges that this solar home faces.
Meachem residence, Raleigh, North Carolina.

The building's metal roof and wood siding reflect the area's heritage. Meachem residence.

Many designers have built successful projects for private clients, but only a few have undertaken the challenge of designing and living in their own creation. As with most learning, there is no substitute for first-hand experience. As living proof of their own design concepts and products, partners John Meachem and Mike Funderburk of Sunshelter have followed this learn-by-doing philosophy. They are neighbors, each with his own solar home in a pleasant rural setting outside of Raleigh, North Carolina. Their homes have metal roofs and wood siding that reflect the historical building style of the area.

The climate in this region requires a certain respect for the winter storms that often result in daytime temperatures well below freezing. But more important is a respect for the high summer temperatures and accompanying humidity. Special design tactics are required to deal successfully with hot, sticky climates. It is easy to warm a house for comfort, but natural cooling is another matter. Meachem and Funderburk have addressed the cooling issue and have come up with a simple yet effective solution.

During a typical summer day, the building is kept closed and the ceiling fan is used to mix the air as much as possible. The fan is adequate, but Meachem plans to increase its speed to improve the cooling effect. This air movement is the primary means of providing comfort, since no dehumidification system is used. At night, when outside temperatures are lower, both cross- and stack ventilation are used

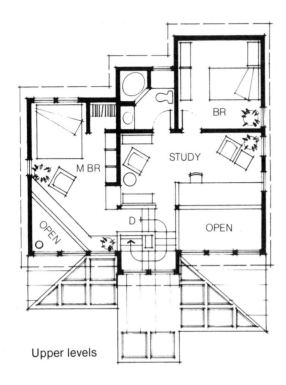

Upper levels

to assist the cooling process. Additionally, seasonal shades are used to protect the sloped glazing from the summer sun.

During the winter, the sun's heat is captured and stored in the concrete and brick floor and in water drums located in the south living spaces. Heated air freely flows to the upper bedroom and study and, as it cools, drops back to the first floor for another cycle. This house reflects the southern tradition of interior transoms and louvers; all doors in the living spaces are louvered, maximizing air movement both winter and summer.

Through the application of common-sense principles and traditional building elements that work, the Meachem house demonstrates contemporary architectural styling and thermal integrity, a refreshing update for comfortable living in the Southeast.

Lower levels

10°

10 FT
3 M

The sun's heat is captured and stored in the south living spaces. Meachem residence.

From the Designer/Inhabitant

We design our buildings to reflect both the historical architectural style of the area and to respond to the client's needs. It is also our goal to heat and cool our designs by passive solar means. In this part of the country, most of the older structures, especially the farm buildings, are built with metal gable roofs. We use them in our designs and have found that they dissipate heat quickly, effectively helping to cool the roof. Also, all our roofs have continuous ridge vents to exhaust warm air from under the metal roof as well as from the interior space. We have found that the combination of ridge venting, interior cross-ventilation, and ceiling fans works surprisingly well to provide cooling during the summer.

During the winter, it is not too difficult to heat passively in this mild climate. We try to minimize glazing for heating purposes because it adds to the summer cooling load as well as to construction cost. Our buildings are designed for the task of cooling rather than heating.

1. winter solar gain
2. radiant thermal storage mass
3. natural convection
4. ceiling fan
5. louvered door
6. summer shade screen
7. low vent windows
8. ridge vent

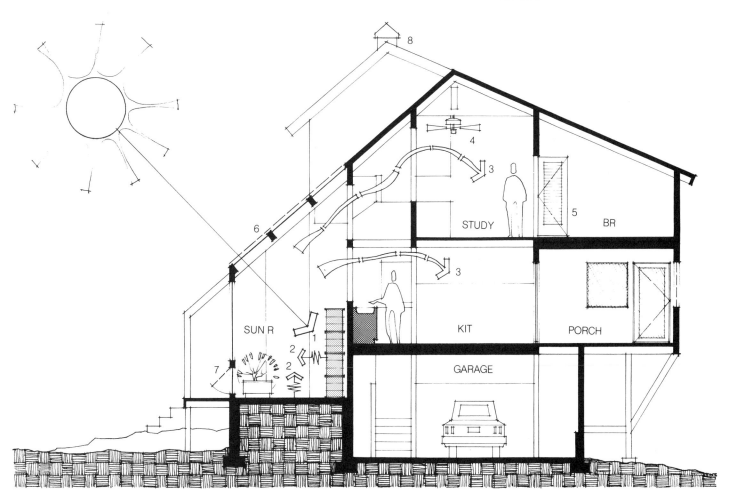

Section

10 FT
3 M

Project Data Summary

Project Information

Project: Meachem residence
              Raleigh, North Carolina
Designer: Sunshelter Design—John Meachem and Mike Funderburk
Builder: Alternative Builders
              Raleigh, North Carolina

Climate Data

Latitude	36.0°N
Elevation	400 FT (122m)
Heating degree days	3,352
Cooling degree days	2,000
Annual percent possible sunshine	61%
January percent possible sunshine	56%
January mean minimum outdoor air temperature	31°F (−.5°C)
January mean maximum outdoor air temperature	52°F (11°C)
July mean minimum outdoor air temperature	68°F (20°C)
July mean maximum outdoor air temperature	88°F (31°C)

Climate features: hot, humid summers; mild winters

Building and System Data

Heated floor area	1,400 FT^2 (130m^2)
Solar glazing area	
Direct gain	500 FT^2 (46.5m^2)
Thermal storage heat capacity	
Tiled concrete floor	4,118 Btu/°F
Water drums	9,330 Btu/°F

PRESS RESIDENCE: Inverness, California

This linear, sculptured house is beaufifully terraced on a hillside. Press
residence, Inverness, California.

All spaces open to the central "interior street." Press residence.

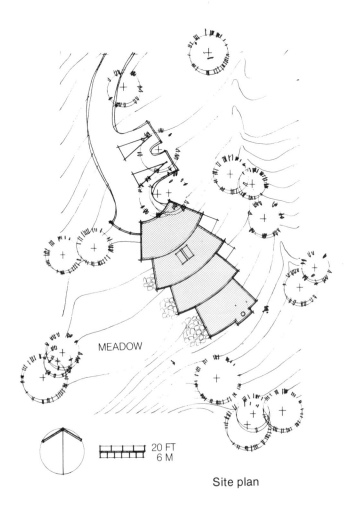

MEADOW

20 FT
6 M

Site plan

The mild climate of the Northern California coast creates something of a challenge for passive solar heating. Accompanying the moderately cool temperatures is a lack of sunshine, abundant summer coastal fog and winter rain. Architect Alex Riley was not inhibited by this, and designed a dynamic passive solar building in this beautiful environment.

The Press residence is an unusual, sculptural house that relates strongly to the site, climate, and local materials. It steps gently down the southern hillside and contains a central "interior street" with access to four levels. The living spaces are on each side of the interior street: the bedrooms and study in the north portion and the kitchen, dining room and living area in the south; each room opens to prevailing breezes as they funnel through the interior street. Although there is little direct sunshine in this climate, diffuse radiation is abundant, and this direct-gain house thrives on it.

The exposed aggregate concrete floor provides the only thermal storage. Backup heating by a hydronic heating grid in the floor provides for, and is an excellent complement to, this passive system. The design expresses the use of native materials well, with exposed fir beams, decking, and exterior cedar siding. The combined effect of the unfolding form, organic materials, and honest structural expression make this home a fine statement of richness through simplicity.

From the Architect

Much of this solution was determined by the desire to preserve the small existing meadow on the site. If we placed the house behind it, the trees to the west would have excessively shaded the house in the afternoon. The solution, a linear arrangement from north to south, maintained the open meadow while permitting solar heating.

The building terraces naturally downhill to the south. This stepping permits a series of clerestories, bringing sun to each space, including the interior street. The fan-shaped plan allows each space to look past one another to the southerly view as well as to obtain solar access. The on-grade slab, with an integrated heating system insulated below, is a successful solution I've been using for several years. This combination allows me to achieve the thermal storage mass needed, and keeps the cost down, since the remainder of the house is wood-framed.

From the Inhabitant

After preaching the value of the sun for many years as a practical, efficient source of energy, it is good to know from first-hand experience that it really works. And it does!

Architecturally, this house is unique: a beautiful, simple design ideally suited for our gently sloping site. We find this an exciting, comfortable house to live in. The decks and patio allow pleasant interaction with the outdoors; the windows and clerestories provide refreshing contact with sky, trees, and birds.

The passive solar system works very well, keeping the house warm in winter and cool in summer. We moved in during the winter, the coldest and wettest period of the year. The first thing we found was that the combination of daytime sun and an evening fire in the fireplace was enough to heat the house comfortably. During colder nights, we simply put on sweaters. The house stays warm even after three days of heavy rain and no direct sun.

In summer, the house is comfortable both night and day, even when the cool summer fog rolls in. This is the first house in this area we have lived in where a fire was not necessary each summer evening. We are able to vent the clerestories to prevent the house from getting too warm on sunny summer days.

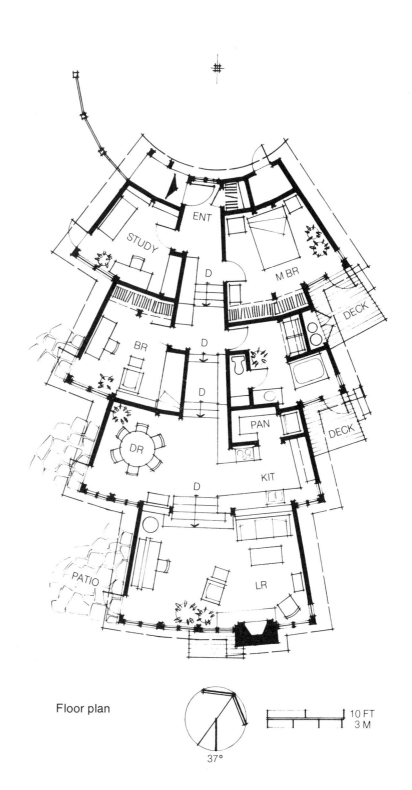

Floor plan

37°

10 FT
3 M

Stepping provides clerestories that heat, light, and ventilate the interior. Press residence.

Best of all, our solar home is a real money saver. Our monthly utility bill is substantially less than in our previous homes with central heating. We used California's solar income tax credit to deduct a major part of the cost of the solar water-heating system. In fact, we were left with substantial tax credits to carry forward to next year. It is a joy to live in such a beautiful home and to live so lightly on the land.

1. winter solar gain
2. radiant thermal storage mass
3. operable clerestory windows
4. solar water heater collector

ENT

"STREET"

10 FT
3 M

LR

Section

Project Data Summary

Project Information

Project: Press residence
Inverness, California
Architect: Alex Riley
Tucson, Arizona
Builder: John Anderson
Inverness, California

Climate Data

Latitude	38.1 °N
Elevation	200 FT (61m)
Heating degree days	3,000
Cooling degree days	0
Annual percent possible sunshine	65%
January percent possible sunshine	52%
January mean minimum outdoor air temperature	42°F (5.5°C)
January mean maximum outdoor air temperature	56°F (13.5°C)
July mean minimum outdoor air temperature	51°F (10.5°C)
July mean maximum outdoor air temperature	64°F (17.5°C)

Climate features: foggy summers, rainy winters

Building and System Data

Heated floor area	1,450 FT² (134.5m²)
Solar glazing area	
Direct gain	290 FT² (27m²)
Thermal storage heat capacity	
Exposed aggregate concrete floor	14,500 Btu/°F

Pole-frame construction solved the problem of building on a flood plain. Abramson residence, Sacramento, California.

Small is beautiful to designer Brent Smith. His concern with the environment led him to formulate a philosophy which dictates efficient use of space, material, and landscape. These three resources are to be used with an uncompromising quality which precludes quantity.

The discipline involved in designing small, efficient homes does not necessarily minimize or reduce construction cost. The emphasis on fine craftsmanship and quality materials makes his designs lasting and enjoyable, a price worth paying. And although the designs are small and necessitate multiple use of space, the quality of living is not lessened. Rather, it is increased through a more sensitive pattern of living.

With this flood-plain site came restraints that many designers would consider too challenging. Smith took vision from this site difficulty by creating a small and efficient house that floats above the landscape. Pole-frame construction proved to be a logical and cost-effective solution to elevating the living structure above the perennial flood plain. A more difficult problem was incorporating thermal storage mass one story up in earthquake country. With proper structural engineering, Smith designed the cantilever pole structure to support the nonstructural concrete floor slab. The poles conveniently extend above the roof, creating support for the shading trellis.

The design of the pole structure, the shading trellis, and the clerestory light monitor are details that, combined with energy efficiency and smallness, make this design especially interesting. It is exciting to see how creative people deal with restrictive design problems to achieve passive solar solutions.

A small, efficient house floats above the lanscape. Abramson residence.

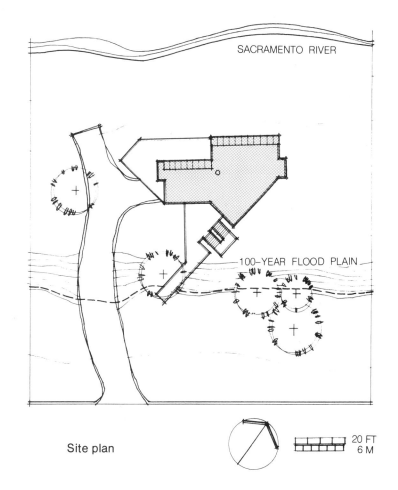

SACRAMENTO RIVER

100–YEAR FLOOD PLAIN

Site plan

20 FT
6 M

From the Designer

Most of my previous work had been in the foothills of the Sierra Nevada, and the sense of place I developed had no bearing on this river site. This design opportunity provided a clean slate and a chance to let a design grow that would be fitting to the flat, agricultural river valley. Two or three times a year, the site is under 2 to 6 feet (.6–2m) of water, and any building must have its first floor at least ten feet above grade.

There were additional restraints: a levee road was along the south of the building location: if the building were to be direct gain, the south-facing glass would be a privacy and security problem. The breezes are also from the south, and since cooling is the major problem in this area, good ventilation is essential. The view of the river is to the northwest and could cause

great discomfort by reflecting the sun during summer afternoons and evenings. Another concern was earthquake design with a raised structure, especially if heavy thermal mass were to be incorporated.

The design uses large light monitors along the north wall to bounce winter sun into the space. This allows solar gain and solved the privacy problem to the south. These and most other windows are insulated with movable shutters. The monitors also act as a solar chimney for summer cooling and ventilation. Fresh air enters the house through an evaporative cooler frame (minus the fan) on the roof. To help with cooling, it was necessary to shade the building as much as possible. This was achieved by continuing the poles up past the roof to support a lightweight trellis. In summer, leafy vine plants intertwine about the structure and guy wires to

shade the building.

The thermal mass is an exposed aggregate concrete floor. Presently, there is not enough thermal mass; we are considering the incorporation of some phase-change thermal storage to increase heat-holding capacity.

The interior is essentially one large space divided by sliding partition walls. The partitions between the dining room and the guestroom–study can be removed and stored. This hides the work desk, making the whole space a large dining area. In a small house it is essential to make use of multiple-use spaces.

An unusual feature of the house is an interior hot tub. With the future installation of a solar water-heating system, the hot tub will be used to heat bodies as well as the house. This has been done for centuries in Japan and, combined with floor-length insulated kimonos, it is a comfortable and enjoyable way to stay warm.

Some of the subconscious images I had when designing were the old industrial piers and marinas, the old river barges and booms, and the riggings of sailboats. I really didn't set out to create a nautical theme, and rather detest that kind of preconceived approach. But I do like to have buildings fit the environment. In my opinion, a lot of solar architecture has no sense of place and, like the typical tract house, could be anywhere.

A major concern of mine is the scale at which we live as a culture. It seems to me that most of our buildings are simply too large. We have tended to equate quality with size. I am interested in working on small, labor-intensive, resource-conserving, finely crafted houses. I feel that we have to reestablish our land use on a planning scale that is more rural. I believe that the single-family residence that passes from generation to generation is the most viable form for American housing.

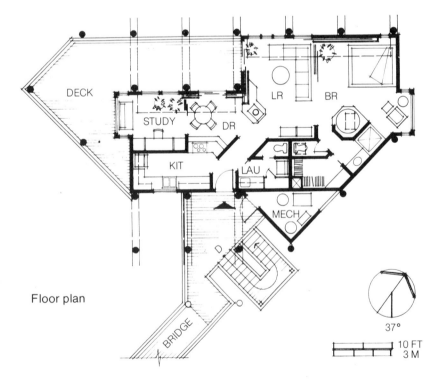

Floor plan

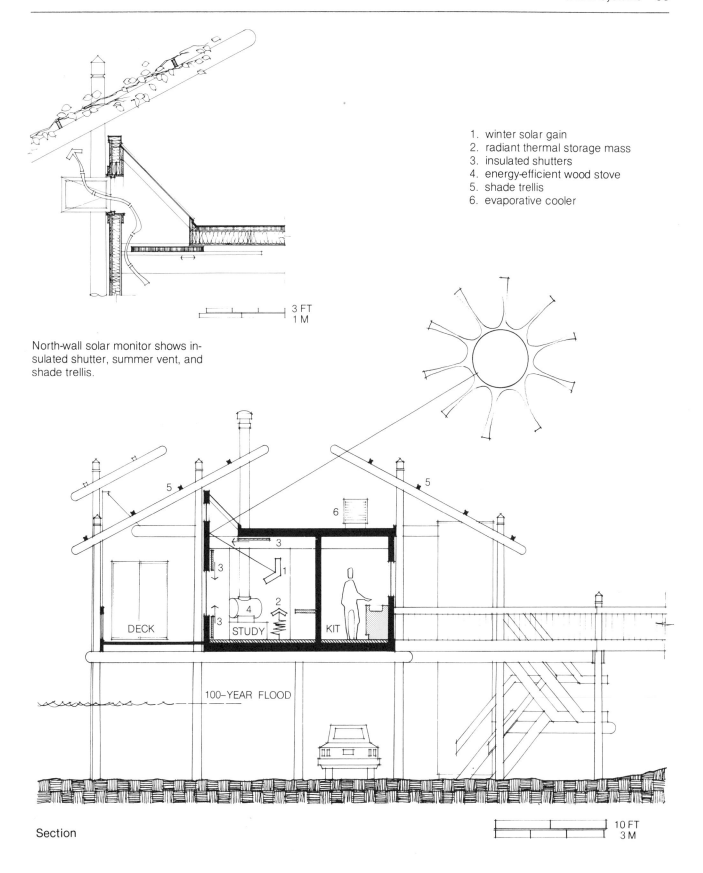

1. winter solar gain
2. radiant thermal storage mass
3. insulated shutters
4. energy-efficient wood stove
5. shade trellis
6. evaporative cooler

3 FT
1 M

North-wall solar monitor shows insulated shutter, summer vent, and
shade trellis.

5

5

6

3

3

1

3

2

DECK

4

STUDY

KIT

100-YEAR FLOOD

10 FT
3 M

Section

Sliding partition walls create multiple-use spaces. Abramson residence.

From the Inhabitant

I feel safe, proud, infinitely rewarded, and joyous in this house. Instead of complicating my life, it has helped me simplify, focus, and balance—goals that emerged as the house evolved. The house has introduced me to a life style I never knew was pertinent. Never one to make much of having friends over, I now enjoy sharing my home with people. Former apartments were often dark when I wished for light, narrow where I wanted space, noisy where I longed for quiet. Compromise was mandatory. What a blessing to live as I now do! The house is small by American standards. It is a maze of corners, a study in how one space can be many. Here, I can change space and light and live efficiently. The house follows my moods and dictates its own needs as the sun and weather change. We are a partnership. The experience of creating and living in this house has reinforced my belief that multiple use and small spaces are necessary to affordable quality. You don't need a room for every activity of your life.

The house looks down on a river and is intended to meld with it. It has forced me outside. From the city culture, I've come to understand the country ethic. The rhetoric of environment is not just babble. This spring, I shall plant my first vegetables, as I planted my first flowers and fruit last year. I am more aware of the seasons and the weather. Part of the experience of this house has been my discovery of how everything interrelates—energy conservation, life style, esthetics. This house has simplified my life, leaving more personal energy for other things.

Project Data Summary

Project Information

Project: Abramson residence
          Sacramento, California
Designer: Brent Smith
          Loomis, California
Builder: Ananda Construction
          Nevada City, California

Climate Data

Latitude	38.8 °N
Elevation	75 FT (23m)
Heating degree days	2,782
Cooling degree days	1,159
Annual percent possible sunshine	78%
January percent possible sunshine	45%
January mean minimum outdoor air temperature	37°F (2.5°C)
January mean maximum outdoor air temperature	53°F (11.5°C)
July mean minimum outdoor air temperature	57°F (14°C)
July mean maximum outdoor air temperature	93°F (34°C)

Climate features: foggy winters; warm summers

Building and System Data

Heated floor area	1,024 FT^2 (95m^2)
Solar glazing area	
Direct gain	164 FT^2 (15m^2)
Thermal storage heat capacity	
Exposed aggregate concrete floor	7,155 Btu/°F

Chapter 3

INDIRECT SYSTEMS

In indirect passive systems, heat is accepted or dissipated at the weatherskin of the building. Typically, collection is separated from the living space, but storage is thermally linked: heat is transferred through a thermal storage mass by conduction, then to the space by radiation and convection. Heat from between the mass wall and the glazing can be conveyed to the interior if the mass wall is vented; such a mass wall is commonly referred to as a Trombe wall, after its inventor. Thermal storage is generally masonry or concrete directly behind the south glazing.

With a water wall, another indirect system, liquid storage is held in barrels or tubes directly behind south-facing glass. During the winter, solar radiation is absorbed by the contained water, which stores it and gradually releases it to the interior. An advantage of using water is its ability to store more thermal energy by volume than masonry. Both mass and water walls should be shaded or vented to the exterior during overheated periods. Additional cooling can be provided by venting the walls at night. A sunspace—a glass-enclosed area—can also indirectly heat the interior, either through conduction or convectively, by openings which allow heated air to flow naturally into the interior.

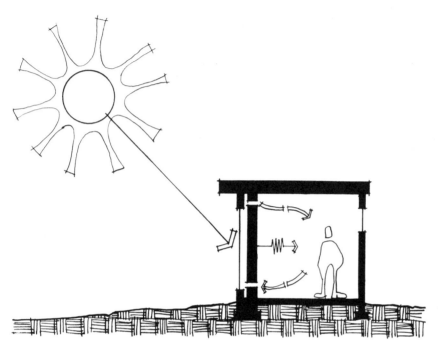

With an indirect-gain system, heat is collected and stored adjacent to the primary living spaces and is thermally linked to them.

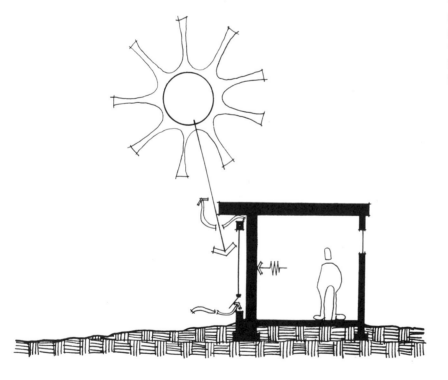

With indirect loss, heat is dissipated at the weatherskin or through intermediate elements; an example is the cooling of mass walls.

The roof pond is an indirect system that places the liquid storage mass at the ceiling. During the heating season, operable insulation panels are moved during the day to expose the storage mass to the sun. Energy is absorbed by the roof pond, and at night the panels are replaced over the storage, allowing the stored energy to radiate to the building's interior. During summer, this process is reversed; internal heat is absorbed by the roof pond, which is insulated above from the high summer sun. At night, the insulation is opened to allow the storage mass to radiate heat back to the sky.

An advantage of indirect systems over direct systems is that the thermal storage acts as a buffer or screen, preventing excessive light, heat, and glare in the living areas. On the other hand, mass walls may lose much of the captured solar heat to the exterior at night if improperly insulated. Certainly, mass or water walls are not ideal if an open view is desired.

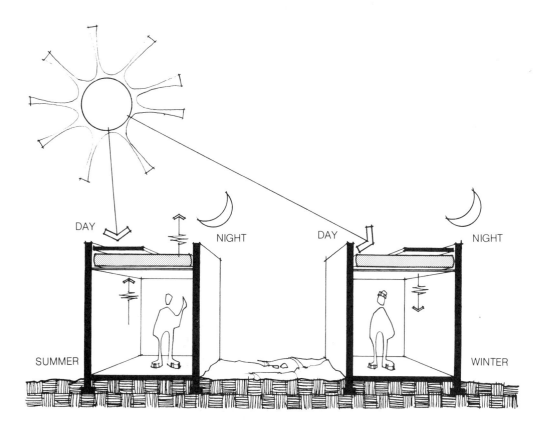

Roof pond

A passive-solar house need not face south. Dewinkel residence retrofit, Madison, Wisconsin.

How does one retrofit a home with no south-facing walls in a fairly typical urban neighborhood? This passive solar home in Madison, Wisconsin, demonstrates one answer. The existing building had a corner edge pointing to the south and fronted on the street, which, unfortunately, runs southwest to northeast. A sunspace on the south corner of the existing building creates the means to collect the sun's heat. Orienting the addition to the south created two additional distinct living spaces: a winter retreat on the west and a summer porch on the east. The porch has no thermal mass and is intended to provide a well-shaded and vented atmosphere. The winter retreat does well with a wood-stove assist, but gets limited solar gain because of its southwest orientation and the low-setting sun. Ironically, shading by an existing tree and a cool floor have made this an excellent summer space as well.

Designed by Don Schramm, Paul Luther, and Mike Utzinger of Prado, the addition has proven successful. The new spaces have improved the view and increased the natural light to the interior. Even with the additional glazing and added square footage, the house stays as cool as it used to during the summer, thanks to cross-ventilation and overhangs. It is heated during winter to the previous comfort standards using approximately 30 percent less fuel.

The neighbors have been intrigued and impressed by the addition. The idea of decreasing heating and cooling bills with an attractive solar addition on a building that doesn't face south is convincing to them, and one nearby neighbor has started a similar remodel.

Flexible awning provides protection from warm-weather overheating. Dewinkel residence retrofit.

From the Architect

The existing home was a compact one-and-a-half-story cube. Its light frame construction would not, at first glance, seem to lend itself to one of the major tenets of passive solar design—massive materials for thermal storage. To further complicate the solar feasibility, no wall faced south.

Our design concepts were easily justified to a household with a history of concern for energy conservation. The owners had watched energy prices rise and had adopted a home energy-conservation program that included daily and seasonal thermostat setbacks, limited heating of low-use spaces (zone control), increased attic and wall insulation, and a supplementary wood-stove heating system. After achieving savings with these measures, they were ready to consider the additional benefits of a passive solar system.

The solar addition was designed with the twofold purpose of increasing living space without increasing energy use. Three separate spaces were desired: a winter retreat, in which to maintain high levels of thermal comfort in Wisconsin's winter, a screened porch for warm-weather activities; and, an attached greenhouse or sunspace, both for in-house, year-round plant growth and heat generation. The basic remodeling problem was to integrate these three spatial and thermal needs into an affordable, easily constructed addition. The design had to blend unobtrusively with the existing structure, floor layout, and surrounding neighborhood, and make use of compatible building materials.

In Wisconsin's winter climate, movable window insulation is vital to year-round utilization of a solar greenhouse. Protection from warm-weather overheating demands flexible shading, since the need for solar penetration into a home is not equal before and after the summer solstice. Very often, solar warmth is needed in April and not in August, even though the sun's altitude is the same.

The solar heart of the addition is the sunspace. It is convectively coupled with the existing dining room and the winter retreat. Both the sunspace and porch serve as buffers between the outdoor chill and interior warmth. Under a planting bench are several water tubes to store heat in the sunspace. An exposed aggregate concrete floor provides additional thermal storage.

The winter retreat is used principally as an evening living area and derives its warmth from three sources: the quarry tile floor, which is heated by direct solar gain; stored heat from the sunspace, vented through registers in the bench or sliding windows above the bench; and a wood stove. Insulating window panels were custom-fabricated by the home owner for use in both the sunspace and winter retreat. Movable insulating window shades are being planned for windows in the older portion of the house.

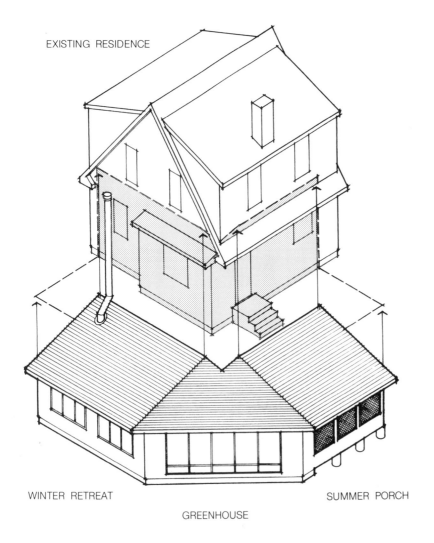

EXISTING RESIDENCE

WINTER RETREAT

SUMMER PORCH

GREENHOUSE

Solar addition integrates with existing southeast and southwest facades.

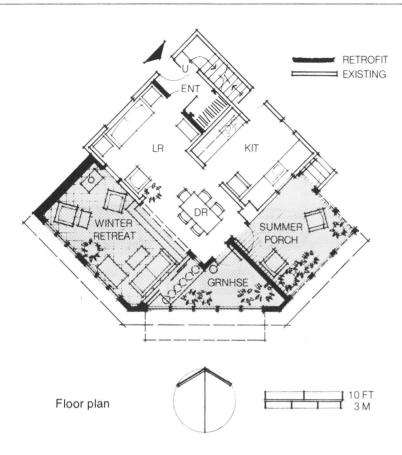

RETROFIT
EXISTING

U
ENT
LR
KIT
DR
WINTER RETREAT
SUMMER PORCH
GRNHSE

Floor plan

10 FT
3 M

Outside shading prevents overheating in warmer months. Fixed overhangs shade both the retreat windows and the sunspace. Deciduous vegetation effectively shades the southwestern windows in the late afternoon. Additionally, a standard canvas awning is used to provide flexible shading at the sunspace. Operable windows allow natural ventilation and are sufficient for cooling needs on most summer days.

Planting bench integrates water tubes and registers to convey heat to winter retreat.

Water tubes beneath planting bench store solar heat. Dewinkel residence retrofit.

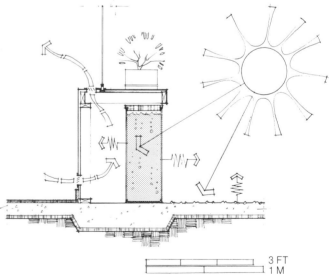

3 FT
1 M

Project Data Summary

Project Information

Project: Dewinkel residence retrofit
        Madison, Wisconsin
Architects: Prado—Don Schramm, Paul Luther, and Mike Utzinger
        Madison, Wisconsin
Builder: Environmental Living, Inc.
        Middletown, Wisconsin

Climate Data

Latitude	43°N
Elevation	858 FT (261.5m)
Heating degree days	7,730
Cooling degree days	460
Annual percent possible sunshine	60%
January percent possible sunshine	47%
January mean minimum outdoor air temperature	8°F (−13.5°C)
January mean maximum outdoor air temperature	25°F (−4°C)
July mean minimum outdoor air temperature	59°F (15°C)
July mean maximum outdoor air temperature	81°F (27°C)
Climate features: very cold winters; warm summers	

Building and System Data

Original floor area	800 FT^2 (74.5m^2)
New floor area	410 FT^2 (38m^2)
Solar glazing area	
Direct gain/sunspace	170 FT^2 (15.5m^2)
Thermal storage heat capacity	
Tiled concrete floor	4,363 Btu/°F
Water tubes	1,176 Btu/°F

Performance Data

Building load factor	8.7 Btu/DAY°F FT^2
Auxiliary energy (heating)	17.2 MMBtu/YR
Auxiliary energy (cooling)	2.4 MMBtu/YR
Solar heating fraction	65%
Night ventilation cooling fraction	81%

WELSH RESIDENCE RETROFIT: Coopersburg, Pennsylvania

An addition to an old cottage made passive-solar heating a new reality. Welsh residence retrofit, Coopersburg, Pennsylvania.

The two-story addition heats not only itself but also one-third of the existing house. Welsh residence retrofit.

As young families grow, so does their need for space. Often, added square footage is only a part of the requirement. Age differences between children as well as different interests require activities to be spatially separated. Seasonal-use spaces are another variable. The desire to expand and contract living areas depending on climate and space-conditioning needs is an interesting possibility for energy-conscious design. It was with these issues in mind that the Welshes hired their friend and architect, Charles Klein, to explore the possibility of an addition to an old but comfortable house. The concern for energy-conscious design, although important, was a second priority.

The result satisfies the space needs, family concerns, and much more. The new family room allows a separate social activity space for adults and children. Even with an increase in square footage, volume, and utility rates, total energy usage has been noticeably reduced. Due to the southern orientation of the existing building, as well as an existing brick facade for thermal mass, a passive solar addition proved a sound, common-sense approach. Inside and out, the home has improved architecturally, providing additional delight for the occupants.

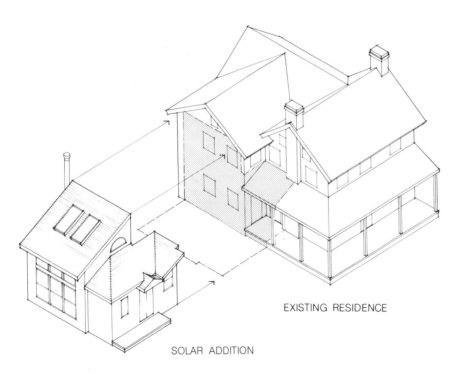

SOLAR ADDITION

EXISTING RESIDENCE

Solar addition integrates with south facade of existing building.

From the Architect

The original house was built around 1830 as a four-room farm cottage. An addition in about 1880 added an extra room on each of the two floors. The Welshes wanted a second living space, an entry foyer, and a first-floor bathroom. The logical location for the addition was on the south side, and the obvious concept was a sun room using the existing brick wall as the thermal mass. Distribution of the heat collected and stored in the sun room to the existing house required a natural convection circuit.

By designing a high ceiling in the new room, I was able to increase the area of the brick wall used for thermal mass. The second floor of the existing house is connected to the sun room, providing a loop for heat transfer. As the air rises and enters the second-floor bedroom, it gives off its heat to the space and then circulates down the stairwell, returning through the kitchen to the sun room to be reheated. The airtight wood-burning stove in the sun room also uses this convective distribution loop to distribute heat throughout the house.

The two-story addition not only heats itself, but one-third of the existing house as well. Interestingly, the extra height in the sun room is a welcome relief to the family, as the rest of the house has low, flat ceilings. This contrast provides a psychological spaciousness which is another asset of the addition, above and beyond its fulfilling the family's program requirements and energy-conservation concerns.

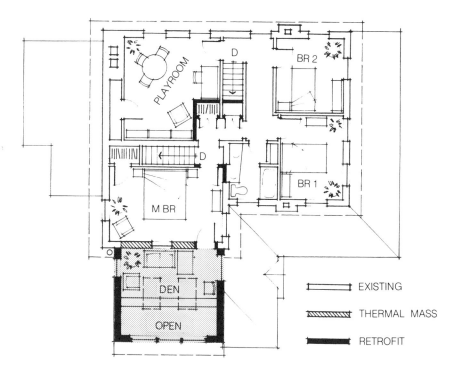

EXISTING

THERMAL MASS

RETROFIT

Upper level

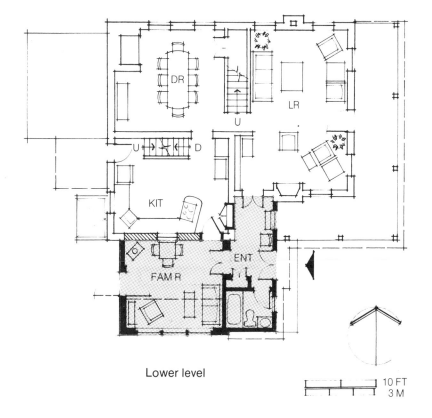

Lower level

10 FT
3 M

Project Data Summary

Project Information

Project: Welsh residence retrofit
                Coopersburg, Pennsylvania
Architect: Charles Klein, AIA
                Bethlehem, Pennsylvania

Climate Data

Latitude	40.5 °N
Elevation	500 FT (152.5m)
Heating degree days	5,810
Cooling degree days	772
Annual percent possible sunshine	55%
January percent possible sunshine	43%
January mean minimum outdoor air temperature	21°F (−6°C)
January mean maximum outdoor air temperature	38°F (3.5°C)
July mean minimum outdoor air temperature	64°F (17.5°C)
July mean maximum outdoor air temperature	86°F (30°C)

Climate features: cold, wet winters; warm, humid summers

Building and System Data

Original floor area	1,900 FT² (176.5m²)
New floor area	400 FT² (37m²)
Solar glazing area	
Sunspace	81 FT² (7.5 m²)
Thermal storage heat capacity	
Brick wall	4,300 Btu/°F

This retrofit shows how typical townhouse characteristics can be used to collect and store solar heat. Metz residence retrofit, Frederick, Maryland.

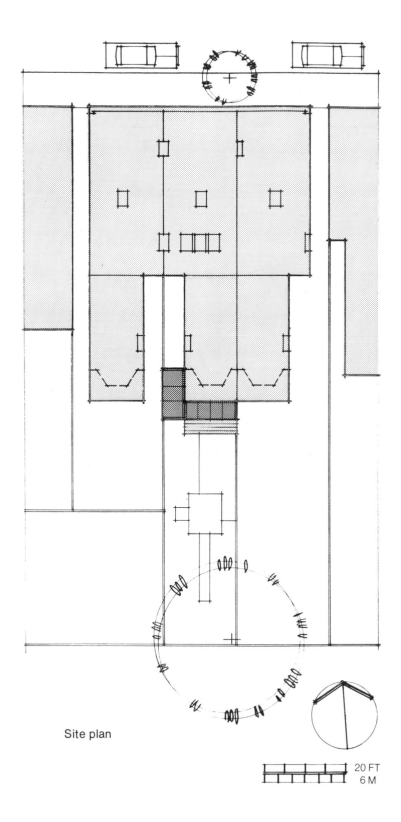

Site plan

20 FT
6 M

The turn of the century urban townhouse is a common and popular home for many city dwellers. Visible in cities across the country, these two- and three-story brick structures are usually well built and much sought after for remodeling. Today, the goal for a family undertaking the task of creating a contemporary habitat in one of these "faces along the street" is making its home energy-efficient. Architect Gene Metz had a double purpose in adding to his home: improving the thermal performance of the uninsulated brick structure, and providing new, bright, sunny, living spaces.

Townhouses usually share one or two side walls with neighbors and are stacked two or three stories high; both factors help conserve energy by reducing the outside surfaces exposed to the elements. The front and rear facades, facing the street and backyard, are normally quite open and uninsulated. This is where most of the natural light and ventilation are admitted to the building.

In this design, Metz took advantage of the south face by using a greenhouse to collect winter sun, and the existing brick mass of the building to store collected heat. The house is carefully designed to admit winter sunlight and to shade itself during the summer. During the heating season, a convective loop circulates warmed air from high in the greenhouse throughout the house and back into the bottom of the greenhouse. Brick floor pavers and water-storage vessels add to the greenhouse's winter heat retention. A ceiling fan assists both summer ventilation and winter heat distribution. Insulative shades cover greenhouse glazing on winter nights.

Care was taken in design to integrate the exterior with the character of this historic building and its neighborhood. Far from compromising it, this passive solar addition complements the beauty of this townhouse and certainly improves its efficiency and comfort.

From the Architect

Designing with passive solar techniques emphasizes one particular set of design options, but not to the exclusion of others. The emphasis on passive concepts requires a high level of discipline. Rather than being restrictive, it tends to provide a richness of design opportunities. It means focusing on providing thermal comfort by natural means and resolving most other design issues from this point of reference. Although it does not mandate a given design direction, it can provide a sense of direction that is compelling. Yet, the broader architectural aspects of "firmness, commodity, and delight" (Vitruvius) need not be sacrificed in the pursuit of energy conservation.

Firmness is inherent in well-built older homes, and brick can store heat. The direct use of sunlight offers distinct advantages in the quest for beauty and delight. Commodity in the form of energy efficiency is the central issue of solar design. A further issue is the creative use of new and conventional materials to satisfy passive design requirements. Delight is the quality of life in a passive solar space.

The addition of a greenhouse to the rear of our ninety-year-old townhouse gave me an opportunity to apply these principles. The character of the old houses is important to us and our neighbors in this historic district. A two-story greenhouse complements the townhouse's vertical proportions. The house's exterior is undisturbed by the solar addition; the glazed facade wraps around it, complementing it in scale, proportion, and color. Existing features were used: the balcony for shading, the doors and windows for heat distribution and ventilation, and the brick for thermal storage. The new space opens off the ground-floor kitchen and the second-level study, providing an eating area, a sitting area, and a space for a large

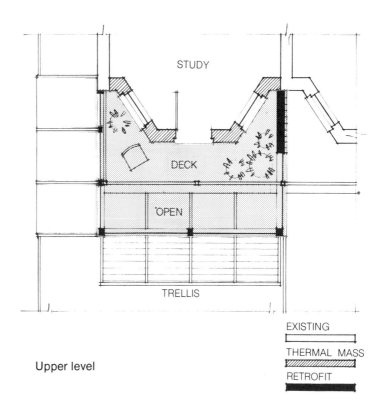

Upper level

EXISTING

THERMAL MASS

RETROFIT

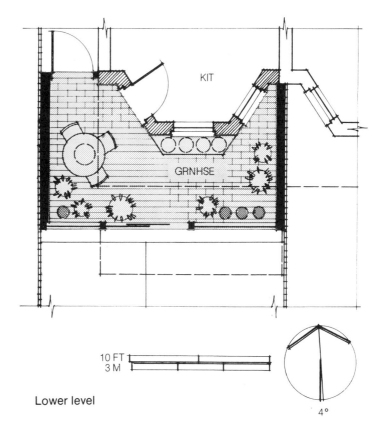

10 FT
3 M

Lower level

4°

variety of plants. New beams give extra support to the balcony, tie the new glazed wall to the existing structure, and project beyond the glazing to frame a trellis for summer shading.

From the Inhabitant

We lived in our 1890 townhouse for two winters before building a passive solar greenhouse. Prior to this addi-tion, the kitchen and the study were uncomfortably cold throughout the winter months. Each room had two large windows and a door; while they let in light and created a pleasant, bright ambiance, they were drafty and the cause of considerable heat loss. When my husband, Gene, designed the two-story greenhouse to extend the kitchen area and encompass the balcony above, it was with a twofold purpose: to provide a proper winter climate for numerous patio plants, and to add warm, comfortable space to the areas of the house in which a great deal of family activity takes place.

One of my favorite pastimes is gourmet cooking, and I grow the various herbs that I use most often. These plants, along with a fairly large collection of tropical and semitropical plants, enhanced the patio and garden during the warm seasons, but had to be brought into the house as winter approached. In addition to expanding the living space, the greenhouse has provided an environment where fresh herbs, scented geraniums, and a variety of other plants grow in profusion. On cold winter days when snow covers the ground, sunlight streams in and we take informal meals surrounded by blooming and aromatic plants. In summer, large sliding doors and windows open, and the greenhouse becomes a pleasant, shaded veranda that leads out to the patio and garden.

The results have been spectacular—a 41 percent reduction in our winter heating bill, and the delightful quality of life in the new spaces.

Greenhouse addition wraps around the existing masonry exterior. Metz residence retrofit.

Opposite page:
Brick floor pavers and water-storage vessels add to the greenhouse's winter heat-retaining capacity. Metz residence retrofit.

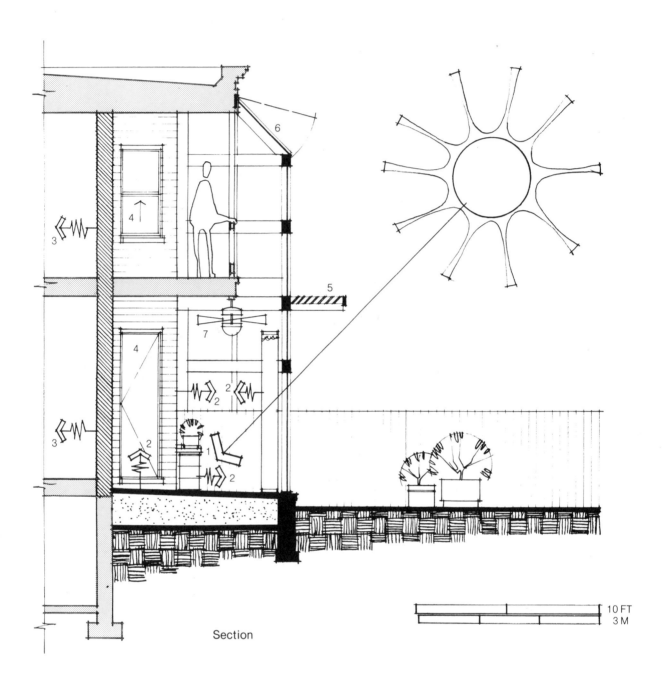

Section

1. winter solar gain
2. radiant thermal storage mass
3. radiant heat to house
4. operable windows
5. shading
6. ventilation
7. overhead fan

Project Data Summary

Project Information

Project: Metz residence retrofit
Frederick, Maryland
Architect/builder: F. Eugene Metz
Oakland, California

Climate Data

Latitude	39.0 °N
Elevation	205 FT (62.5m)
Heating degree days	5,087
Cooling degree days	1,100
Annual percent possible sunshine	60%
January percent possible sunshine	50%
January mean minimum outdoor air temperature	29°F (−1.5°C)
January mean maximum outdoor air temperature	48°F (9°C)
July mean minimum outdoor air temperature	67°F (19.5°C)
July mean maximum outdoor air temperature	86°F (30°C)

Climate features: cold winters; humid summers

Building and System Data

Original floor area	2,450 FT² (227.5m²)
New floor area	150 FT² (14m²)
Solar glazing area	
Sunspace	365 FT² (34m²)
Thermal storage heat capacity	
Brick pavers over concrete slab	2,250 Btu/°F
Brick walls	9,027 Btu/°F
Water tubes	2,586 Btu/°F

A house within a house: the concrete-block inner house is encased in an insulated weatherskin. Roberts residence (Sundance), Reston, Virginia.

"A house within a house" is the underlying concept for this residence, located in Reston, Virginia. The "outer house" or shell is wood-clad insulated weatherskin that virtually encases the concrete-block "inner house." At a cost competitive with conventional buildings in the area, architect Walter F. Roberts, Jr., was able to create this double shell. The outer shell, which derives its form from local site and climate characteristics, recalls the region's traditional architecture. The inner shell expresses the pure geometry of masonry construction. This masonry core offers thermal storage mass which is directly linked to all living spaces and helps to moderate interior temperature swings. The massive thermal storage wall adjacent to the south glazing is heated directly by the sun. Excess heated air from the space between is drawn by a fan to the masonry walls in the northern portion of the building. This distribution loop encourages balanced heat throughout the building.

This house offers something old and something new. It serves as an example of a successful design which relies on historic vernacular architecture to derive its form, while at the same time incorporating new materials and concepts to improve efficiency within its function.

Masonry creates a protective space that is secure and enduring. Roberts residence (Sundance).

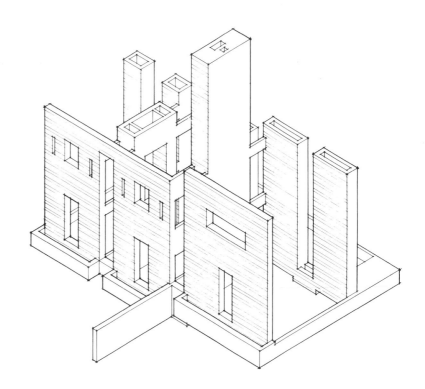

Concrete masonry, inner house

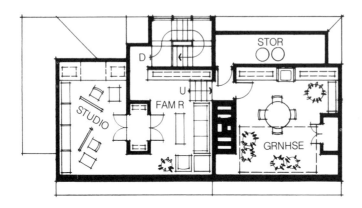

Upper levels

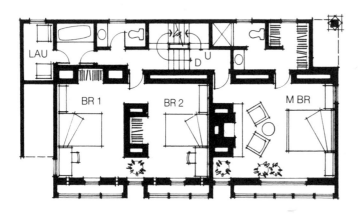

Mid levels

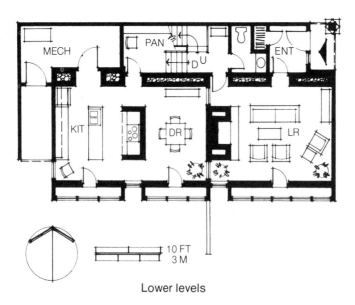

Lower levels

From the Architect/Inhabitant

The concept for Sundance was to develop an indirect passive solar system with heat stored in mass walls enclosing all living spaces. There were a number of reasons why this idea was chosen. The indirect system allows heat to be absorbed and stored with a minimum of overheating of the living spaces. Storing heat in thermal mass directly adjacent to living spaces eliminates reliance on auxiliary heat from isolated storage. This system also allows direct radiant heat to be emitted from two sides of most living spaces.

The development of the house-within-a-house concept evolved into two very distinct house forms. The inner house, which acts as the thermal storage medium, had to be massive. Heavyweight concrete block with sand-filled cores was chosen for reasons of economy and esthetics. This masonry form creates the feeling of a protective space both secure and enduring. The outer house, on the other hand, was by function to act as a thermal cocoon. Heavily insulated walls, sheathed in wood, shield the inner house from the elements and temperature extremes. Dominated by three chimneys, Sundance's roof recalls the form of eighteenth- and nineteenth-century homes, in which these roof structures also served as flues and ventilating chimneys.

Operationally, the house is controlled by opening the cocoon or outer house to allow the penetration of sunshine or cool breezes, depending on the season. When sunlight is available during the winter months, the insulated curtains which cover the entire south side of the house automatically open. The sun then penetrates and begins to warm the inner mass. As the south mass wall is heated, a fan-assisted convective air loop drives the warmed air up through the attic spaces and down to the north mass walls, where heat is stored. The air then returns through the crawl space to the south wall. As darkness falls, so do the thermal curtains, preventing the escape of heat to the outside and allowing the warmth to radiate into the house.

During summer months, the insulated curtains are closed during the day, preventing the sun from penetrating and warming the inner mass. The windows are also closed, preventing the infiltration of hot, humid air. With the windows closed, the only air that can enter the house is through "coolth tubes"—pipes buried in the earth. As the two thermal chimneys located at the roof's peak exhaust warm interior air, exterior air is precooled by the earth as it is drawn through the tubes. The precooled air circulates along the south mass wall, cooling it. In the evening, when the outside air temperature drops, the curtains and windows are opened and fresh air is drawn through the living spaces up the stairwell to be exhausted at the chimneys. This process flushes the inner house of any heat built up during the day.

Because the house works within natural laws, it has instigated a return to nature. It has awakened an awareness of our environment and created a sense of oneness with it. It has rejuvenated the inner joy of self-sufficiency that can still be achieved within our culture.

1. winter solar gain
2. radiant thermal storage mass
3. air-circulation loop
4. fan
5. insulative shades

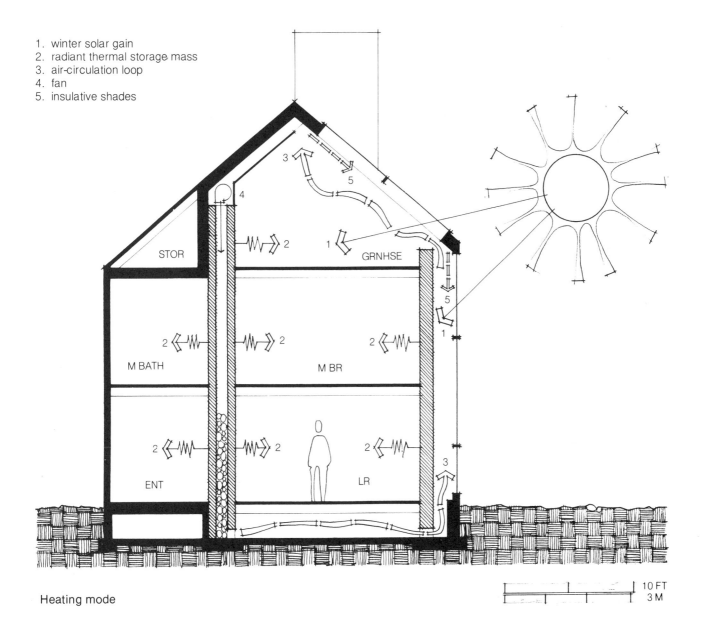

Heating mode

10 FT
3 M

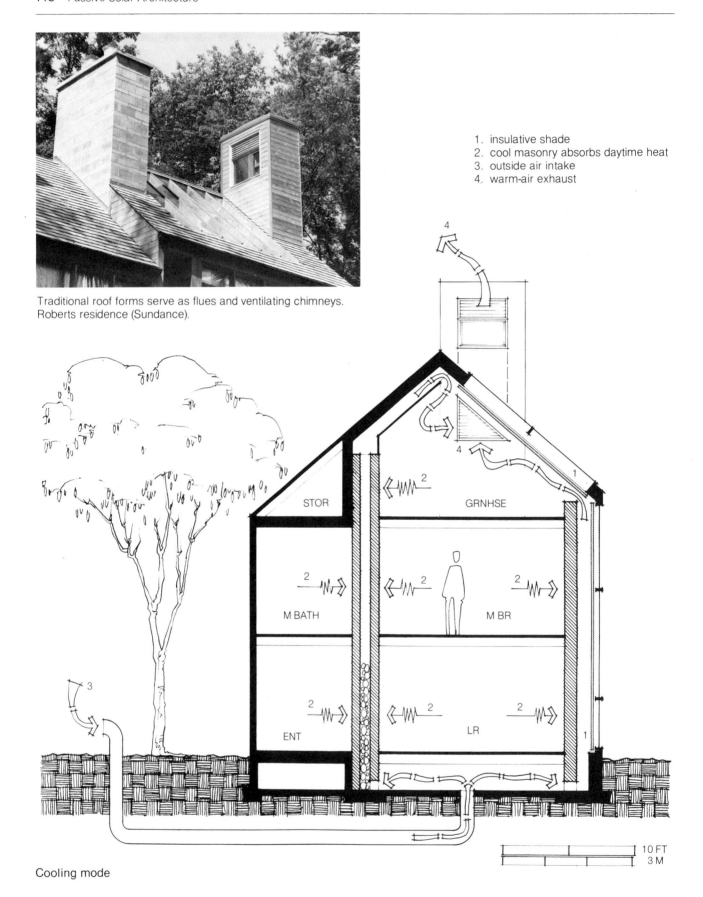

Traditional roof forms serve as flues and ventilating chimneys.
Roberts residence (Sundance).

1. insulative shade
2. cool masonry absorbs daytime heat
3. outside air intake
4. warm-air exhaust

STOR

GRNHSE

M BATH

M BR

ENT

LR

10 FT
3 M

Cooling mode

Project Data Summary

Project Information

Project: Roberts residence (Sundance)
Reston, Virginia
Architect/builder: Walter F. Roberts, Jr., AIA
Reston, Virginia

Climate Data

Latitude	39.0 °N
Elevation	332 FT (101m)
Heating degree days	4,962
Cooling degree days	874
Annual percent possible sunshine	55%
January percent possible sunshine	43%
January mean minimum outdoor air temperature	23°F (−5°C)
January mean maximum outdoor air temperature	41°F (5°C)
July mean minimum outdoor air temperature	64°F (17.5°C)
July mean maximum outdoor air temperature	86°F (30°C)

Climate features: hot, humid summers, cool and cloudy winters

Building and System Data

Heated floor area	2,236 FT^2 (207.5m^2)
Solar glazing area	
Indirect-gain mass wall	592 FT^2 (55m^2)
Sunspace glass	144 FT^2 (13.5m^2)
Thermal storage heat capacity	
Indirect-gain mass wall	14,820 Btu/°F
Interior masonry wall (with air flow)	23,440 Btu/°F
Interior masonry wall	11,560 Btu/°F

Performance Data

Building load factor	6.1 Btu/DAY°F FT^2
Auxiliary energy (heating)	2.2 MMBtu/YR
Auxiliary energy (cooling)	0.4 MMBtu/YR
Solar heating fraction	94%
Night ventilation cooling fraction	94%

The Tennessee Valley Authority sponsors the building of solar homes as part of its energy-conservation program. TVA Solar Residence 5, Glasgow, Kentucky.

Many public and quasi-public agencies, from the federal to local level, have implemented energy-conscious programs to popularize solar architecture. The methods range from design-award construction grants to fully financed demonstration programs. One of the most successful and dynamic of these programs has been developed by the Tennessee Valley Authority, a federal corporation that provides electricity for portions of Alabama, Kentucky, Tennessee, and Mississippi. With the advent of the energy crisis, energy conservation is as important as power generation to assure an adequate electrical supply. Many utility companies have begun to understand the necessity of reducing unlimited demand on their capabilities by implementing energy-conservation programs. TVA became involved in passive solar energy in 1977 with three home designs. After receiving an enthusiastic response to these designs, it established the Solar Homes for the Valley project.

Several home designs have now been developed, and demonstration houses have been built throughout the service districts. The Solar Applications Branch has established a large in-house group of solar architects and technicians to create designs and programs to disseminate solar technology. It has also tapped the national pool of solar expertise to educate its staff and to produce design guidelines and workbooks for general use. Its primary goals are:

1. To educate home buyers to the fact that passive solar homes are cost-effective and desirable
2. To convince financers that passive solar homes are a good investment
3. To demonstrate to builders that passive solar homes are easy to build, maintain, and market

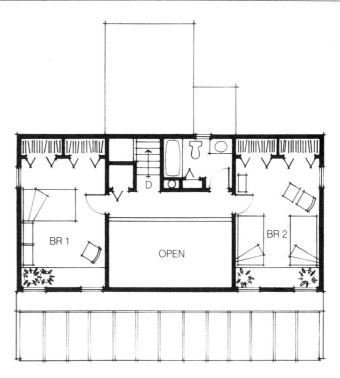

Upper level

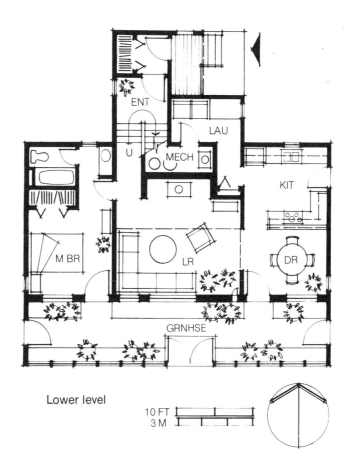

Lower level

10 FT
3 M

Dark-colored mass wall in the greenhouse collects and stores the sun's heat.
TVA Solar Residence 5.

From the Architect

A primary goal of Solar Residence 5 was middle-income acceptance. In the past, solar principles have been largely applied to expensive, custom-built single-family residences. It is important that the lessons learned from these solar homes be translated into prototypical solar design for commercial home building. This demonstration project not only introduces solar to the general public, but generates new designs which are much more interesting than the usual product.

1. winter solar gain
2. radiant thermal storage mass
3. energy-efficient wood stove
4. greenhouse vent/fan
5. warm-air exhaust

Important programming features were the incorporation of conventional construction practices, adaptability to flat or gently sloping sites in suburban or rural areas, and family self-sufficiency. Finally, the home was to contribute to a more relaxed atmosphere for living in our fast-paced world. The use of familiar forms that symbolized "home" in the area was a starting point.

It was desirable that the solar system contribute more than physical comfort. The greenhouse was a natural choice. It was familiar to the public in form and function and required no difficult or untried construction. When handled sensitively, a greenhouse can enhance a dwelling architecturally by offering a changing, pleasurable environment. By providing a food-production capability, each family can become more self-reliant.

Simple passive design was favored over more complicated active solar systems, which were initially explored. A dark-colored mass wall in the greenhouse collects and stores the sun's heat. This heat is indirectly radiated to the living spaces; the doors and a tall, narrow vent can be opened to allow direct convective heat flow. In summer, air is drawn throughout the house and exhausted at high vents. The greenhouse is power-vented by a thermostatically controlled fan to prevent overheating.

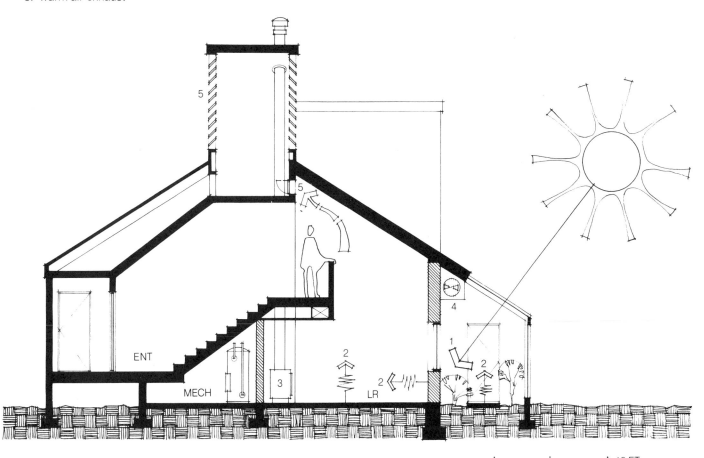

Section

10 FT
3 M

Project Data Summary

Project Information

Project: TVA Solar Residence 5
    Glasgow, Kentucky
Architect: W. Adkins
Designer: Danny Brewer
Builder: Biggers and Armstrong Construction Co.
    Glasgow, Kentucky

Climate Data

Latitude	37.0 °N
Elevation	536 FT (280m)
Heating degree days	4,219
Cooling degree days	1,467
Annual percent possible sunshine	57%
January percent possible sunshine	41%
January mean minimum outdoor air temperature	28°F (−2°C)
January mean maximum outdoor air temperature	48°F (9°C)
July mean minimum outdoor air temperature	66°F (19°C)
July mean maximum outdoor air temperature	90°F (32°C)

Climate features: cool to cold winters; humid summers

Building and System Data

Heated floor area	1,552 FT^2 (144m^2)
Solar glazing area	
Direct gain	157 FT^2 (14.5m^2)
Greenhouse	456 FT^2 (42.5m^2)
Thermal storage heat capacity	
Dark-colored concrete floor	8,580 Btu/°F
Concrete masonry, Trombe wall	10,550 Btu/°F
Wood stove, masonry wall	1,610 Btu/°F
Brick and earth floor in greenhouse	1,600 Btu/°F

SUNSTONE: Phoenix, Arizona

"Waterbeds" on the roof bring solar cooling to the desert. Sunstone, Phoenix,
Arizona.

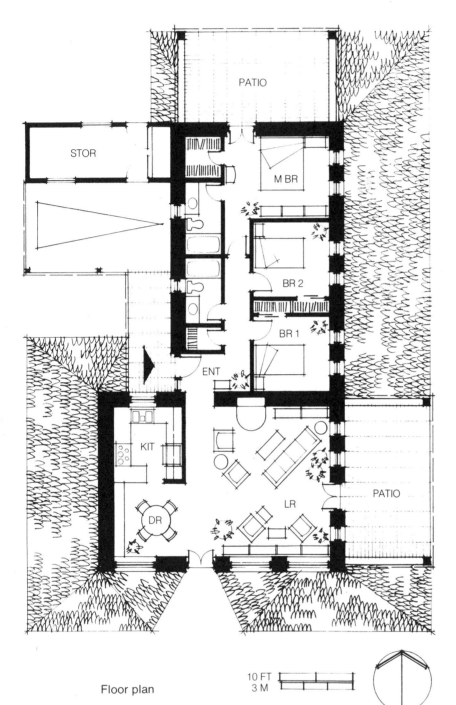

Floor plan

10 FT
3 M

Designing a passive solar house for a hot arid region presents a number of unique challenges. First, and most obvious, is the need for well-shaded and thermally protected spaces. Second is the need for heat rejection or dissipation. Third, ventilation and air motion are needed to achieve comfort. When used effectively, either independently or in combination, these elements can maintain pleasing temperatures throughout the year. Deserts are known for large daily temperature swings. Even when temperatures exceed 100°F (37.5°C) on a summer day, they can dip into the 60s (16–18°C) at night. Generally, clear skies accompany these nighttime lows.

A concept that works well with clear night skies was tested in the 1960s by Harold Hay. He found that water in a simple ice chest could be kept at a nearly constant temperature with little effort, even in the desert. Simply explained: during summer days, an insulating lid is placed over the water-filled chest to keep it relatively cool. At night, the lid is removed and the water is cooled by radiating its heat to the clear sky, commonly referred to as night-sky radiation. During winter, the opposite occurs; the lid was closed at night, allowing the water to retain the heat collected during the day. Since that time, this system—called Skytherm—has been tested in several climate zones and has proven to be effective for both heating and cooling buildings.

Hay and Daniel Peter Aiello of Janus Associates have worked jointly to develop a modified "roof pond" prototype that solves some of the problems of structural loading and the gearing and tracking of the movable insulation. With Sunstone, architect Aiello and Guilford Rand have come up with a design that works well in its desert environment. High-density concrete walls, totaling two feet in thickness, are arranged with a 3-inch (7.5cm) internal insulating core. A highly textured exterior surface reflects and shades itself from desert sunlight, decreasing the cooling load of the building. All door and window openings are deeply recessed for maximum shading. The windows are also placed high in most rooms to help diffuse daylight across the ceilings.

The building integrates energy considerations with pleasant living. In this microclimate, daytime cross-ventilation and solar gain are allowed in the living and dining area, which is oriented east-west. For nighttime cross-ventilation and morning winter sunlight, the bedrooms are oriented north-south. Additionally, they are buffered from the western afternoon sun by the carport.

Sunstone is an insulated, massive, well-shaded, and ventilated building. The roof acts as a dynamic thermal balance to the seasons. Combined, these strategies create a well-tempered desert environment, one which works beautifully with the patterns of nature.

Top view of roof pond

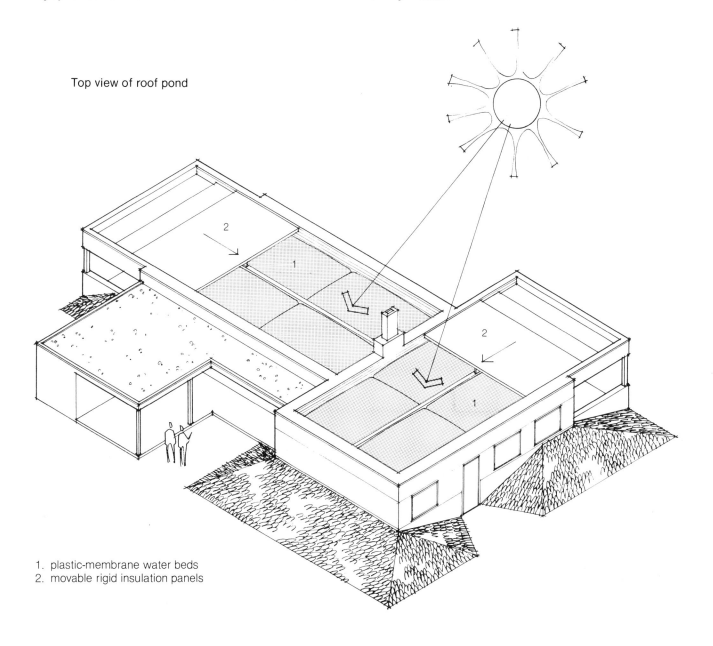

1. plastic-membrane water beds
2. movable rigid insulation panels

From the Architect

This house is an evolved example of passive solar design for the arid Southwest. The attributes of concrete construction to minimize climatic effects and provide thermal stability are coupled with the cooling and heating capabilities of a solar roof pond. The climatic impacts on the building are significantly reduced by a heavily textured and massive concrete surface. The exposed aggregate surface acts as tiny self-shading pockets which also refract and scatter light away from the building. The textured surface, with its pockets and voids, also sets up a fine insulating air layer. Testing has shown up to a 5°F (2.8°C) lower temperature between this particular wall and a similar one with a smooth surface.

The wall itself is two feet (61cm) thick. The external exposed aggregate layer, three inches (7.5cm) thick, acts as a skin buffer. Behind this is three inches (7.5cm) of rigid insulation, which forms a continuous thermal barrier around the building. This layering keeps summer heat out and winter heat in. The remaining eighteen inches (45.5cm) of concrete at the building's interior acts as thermal mass, storing coolness in the summer and warmth in the winter.

The roof pond consists of plastic-membrane "thermoponds" lying on a corrugated steel deck. This acts as the building's roof and ceiling. Above the thermoponds are movable rigid insulation panels which can be manipulated by hand or by a small motor. During summer, the panels protect the water from the intense outside heat. The water, remaining cool, absorbs internal heat, thereby keeping the living spaces cool. At nightfall, the panels are opened and collected heat is dissipated to the night sky by radiation, and to the atmosphere by convection and evaporation. Thus cooled, the thermoponds are ready to begin the cycle again at daybreak and the panels are closed. In winter the conditions and operation are reversed. The panels are opened during the day and the pond gathers solar energy to heat the building. At night, the panels are closed to prevent heat loss.

The coupling of the roof pond with the thermal storage of concrete construction results in an arid-region structure that is economical to construct, energy efficient, and longlasting. Sunstone has maintained a constant 74–76°F (23.5–24.5°C) interior temperature in all seasons. The cost to cool and heat the house is estimated to be no more than $25 a year, in an area where cooling bills can easily approach $100 and more a month.

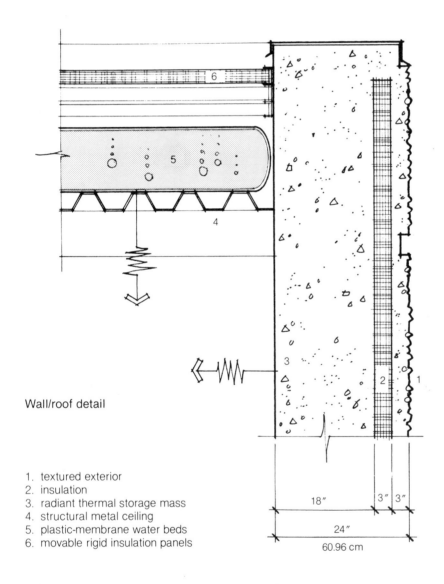

Wall/roof detail

1. textured exterior
2. insulation
3. radiant thermal storage mass
4. structural metal ceiling
5. plastic-membrane water beds
6. movable rigid insulation panels

18″   3″ 3″
24″
60.96 cm

Project Data Summary

Project Information

Project: Sunstone
       Phoenix, Arizona
Architect: Daniel Aiello with Guilford Rand, Janus Associates, Inc.
       Tempe, Arizona
Builder: Stanley Construction Co. and Advanced Building Systems, Inc.
       Phoenix, Arizona

Climate Data

Latitude	33.5 °N
Elevation	1,117 FT (340.5m)
Heating degree days	1,765
Cooling degree days	4,000
Annual percent possible sunshine	86%
January percent possible sunshine	77%
January mean minimum outdoor air temperature	38°F (3.5°C)
January mean maximum outdoor air temperature	65°F (18.5°C)
July mean minimum outdoor air temperature	78°F (25.5°C)
July mean maximum outdoor air temperature	105°F (40.5°C)

Climate features: hot, arid summers with large daily temperature swings

Building and System Data

Heated floor area	1,800 FT² (167m²)
Solar glazing area	
Roof pond	1,280 FT² (119m²)
Thermal storage heat capacity	
Tiled concrete floor	18,000 Btu/°F
Roof pond	73,220 Btu/°F
Exterior concrete walls	61,200 Btu/°F

Chapter 4

ISOLATED SYSTEMS

In isolated passive systems, collection or dissipation is adjacent to or apart from the weatherskin and remote from the primary living spaces. This allows the solar system to function somewhat independently of the building interior, although heat may be drawn directly from the thermal storage mass as needed.

Thermosiphon or natural convection is one type of isolated gain system. It relies on the natural rise and fall of a fluid, such as air, as it is heated and cooled. As the sun warms a collection surface, the air rises, simultaneously pulling cooler air from the bottom of the storage and causing a natural convection loop. Collected heat can be conveyed to the interior space for immediate heating or to thermal storage mass for later use. The thermal storage can either be isolated from or integral with the living space. Once the heat is conveyed to the interior space or storage, the air falls and returns to the collection area, beginning another cycle.

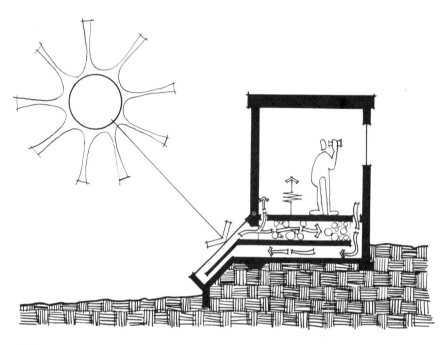

With an isolated-gain system, heat is collected adjacent to or apart from the weatherskin and stored either apart from or in the living space.

Many variations to this system are possible. For example, the high point of the convection loop can be opened in summer to allow the heated air to escape, thus inducing precooled air through the interior spaces or the storage mass. As in other systems, the potential exists for summer overheating of the living spaces.

The double envelope is another type of isolated gain approach. Solar collection in a greenhouse or sunspace heats air that then flows around the building's interior core. Thus, a protective layering of tempered air helps insulate the house.

Two distinct advantages of isolated gain are that only the collector, not the total building, need face the sun, allowing more architectural freedom; and the thermal storage mass can be charged without necessarily affecting interior temperature. Heat can be transferred to the living spaces when desired, allowing more control and reducing the potential for overheating.

With isolated loss, heat is dissipated away from the weatherskin; for example, induced air can be precooled in the earth's mass.

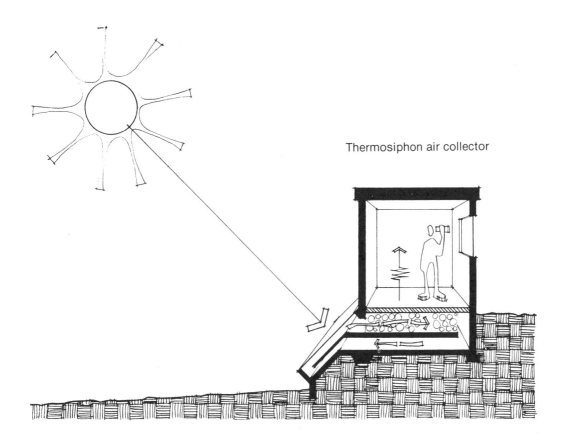

Thermosiphon air collector

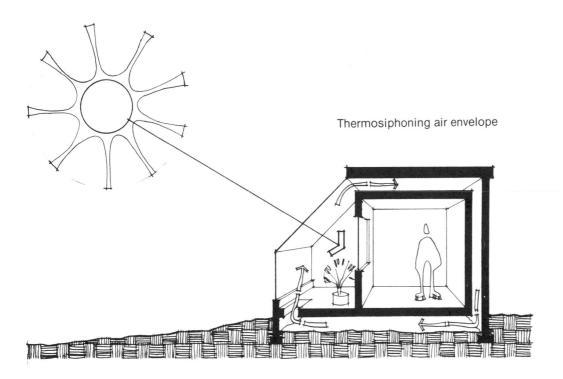

Thermosiphoning air envelope

Remote collectors do not compromise either architecture or siting. La Tierra residence, Santa Fe, New Mexico.

The high, cool, and dry Sangre de Cristo Mountains of northern New Mexico have been the birthplace of many solar projects. In this mecca of passive solar architecture, many innovators developed their concepts, and work continues on all levels, from grass roots to technically advanced. The popularity of passive solar techniques has affected all buildings in the area—banks, condominiums, medical buildings, religious retreats—you name it, and someone has probably built one using passive solar concepts. It has been said that 90 percent of the new buildings around the city of Taos utilize solar energy of some type.

Modern pueblo style for an up-to-the-minute solar home. La Tierra residence.

Floor plan

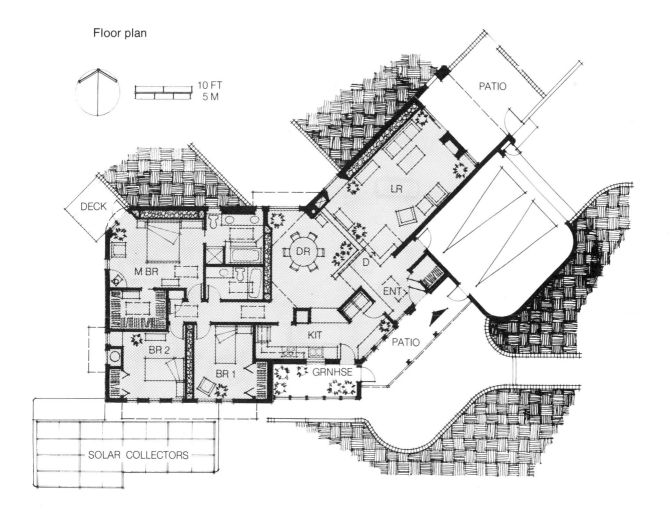

10 FT
5 M

PATIO

DECK

LR

DR

D

M BR

ENT

BR 2

KIT

PATIO

BR 1

GRNHSE

SOLAR COLLECTORS

Thermal storage walls and floors act as large radiators to slowly and evenly radiate heat. La Tierra residence.

The popular pueblo style native to northern New Mexico is a fine example of indigenous design techniques that are well suited to their environment. The Pueblo Indians built massive adobe structures which were responsive to seasonal solar patterns. These sturdy, functional buildings have such a high volume of thermal mass that they tend to remain cooler in summer and warmer in winter than average ambient air temperatures. With the emergence of the energy crisis, south-oriented double glazing

and exterior insulation were added to these traditional forms, creating more effective direct-gain buildings—a small but significant step in energy-conscious design.

Today, the innovative work continues in the Southwest. Architect Mark Jones, inspired by the work of Scott Morris and Steve Baer in natural convection (thermosiphon) air-collection systems, has designed a series of isolated-gain passive solar residences. The La Tierra house typifies the modern pueblo style of the area,

and is comparable in cost and flair to similar well-designed custom-built homes. The unique difference is that it is primarily heated by the silent, subtle flow of warm air from remote collectors to several internal vertical rock-bed walls. Unobtrusive—they appear to be ordinary interior walls—these thermal storage walls act as large radiators, slowly and evenly distributing their stored heat to the inside. Although the building is in adobe style, the exterior wall is made of 2-by-8-inch (5-by-20.5cm) frame walls, well insulated and stuccoed to achieve the adobe look.

The remote or isolated collection of this thermosiphon system allows placement of the solar collection surface without compromising the architecture's orientation, fenestration, thermal mass, or views. This design also incorporates the domestic solar water-heater collectors into the space-heating collector array.

Thermosiphon systems have great potential for space heating, cooling, and ventilation. One limiting factor is that the collector must be located level with or below the thermal storage to allow natural convection. Thermosiphon heat flow could be well suited to high-rise or other high-density structures.

From the Architect

Although I have done projects using Trombe walls, direct gain, and greenhouse applications, the majority of my recent work has involved air-siphon systems. Our group has now completed several medium-sized houses of this type. As far as I know, this comprises the majority of serious full-scale air-siphon projects built. I don't know why more people haven't done work of this type, perhaps because air siphoning has not received nearly the publicity of other passive approaches and is not widely understood. One limitation is that a really effective siphon system needs a south-sloping site. Another is that since air siphons operate at very low

velocities and with very low pressure differentials to drive air through the storage media, they require some fairly careful calculations of a type not covered by most conventional engineering charts. It may be that some people don't even consider air siphons a passive technique. I strongly disagree. None of our air-siphon buildings has a fan larger than one-third horsepower, none has fans on the collector/storage loop, and the most recent project utilizes no fans in any part of the system.

Why use air siphons at all? The use of a collection/storage loop separate from the space itself tends to minimize temperature fluctuations in the space, eliminates any glare and fading problems, and offers more flexibility in

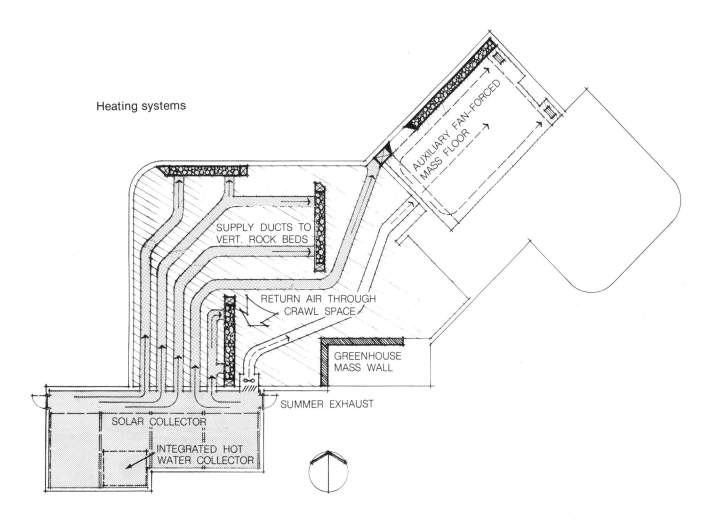

Heating systems

AUXILIARY FAN-FORCED MASS FLOOR

SUPPLY DUCTS TO VERT. ROCK BEDS

RETURN AIR THROUGH CRAWL SPACE

GREENHOUSE MASS WALL

SUMMER EXHAUST

SOLAR COLLECTOR

INTEGRATED HOT WATER COLLECTOR

the organization and orientation of spaces, including the handling of views. For example, the site on which this project is built has no south views, but has excellent northwest and northeast views.

Our air-siphon projects have evolved considerably over the last couple of years. The earliest used an air-to-rock siphon, with fan-forced hot-air distribution to the living space. The second generation used a combination of fan-forced air distribution, direct radiation through the rock-bin walls to the living space, and convective air flow from the rock bins to the living space. This third-generation project uses dispersed rock bins in tall, narrow wall configurations, with heat distribution to the living space only by radiation from the thermal storage walls.

A word about moving parts: we have been working to reduce the mechanical components to the simplest and fewest possible. With the elimination of fans, all that is left is dampering to prevent reverse heat flow (back siphoning) when the sun is not shining. We have found that these dampering installations must be very tight, for even small cracks can result in substantial losses from rock storage. Really tight dampers are hard to purchase and hard to build.

One feature of the project is that the rock storage walls are entirely encased within the insulated building envelope, greatly reducing nocturnal heat loss to the outside as compared to a Trombe or mass wall, and allows the designer to place the heat source where it is most needed. I don't feel that a passive design should necessarily be pure and use only one system when combining different techniques in the same building is reasonable. For example, this building, which is predominately air siphon, also uses a greenhouse/mass wall and direct gain to some degree.

Collector/storage wall detail

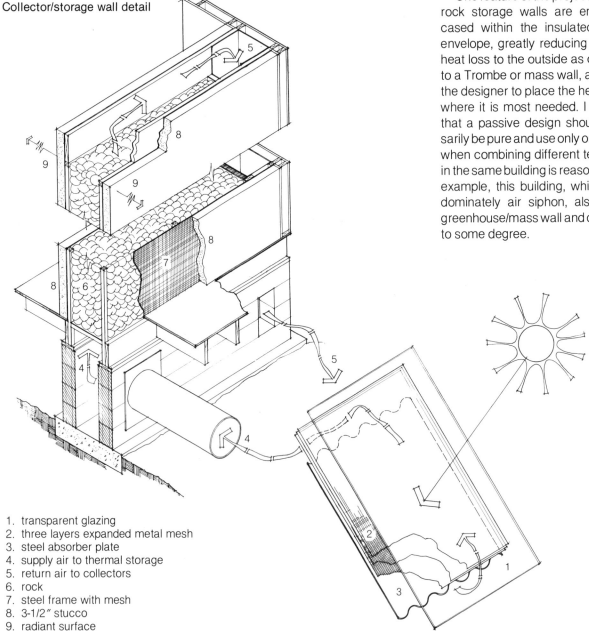

1. transparent glazing
2. three layers expanded metal mesh
3. steel absorber plate
4. supply air to thermal storage
5. return air to collectors
6. rock
7. steel frame with mesh
8. 3-1/2″ stucco
9. radiant surface

The north entry is characteristic of this building, so well suited to its environment. La Tierra residence.

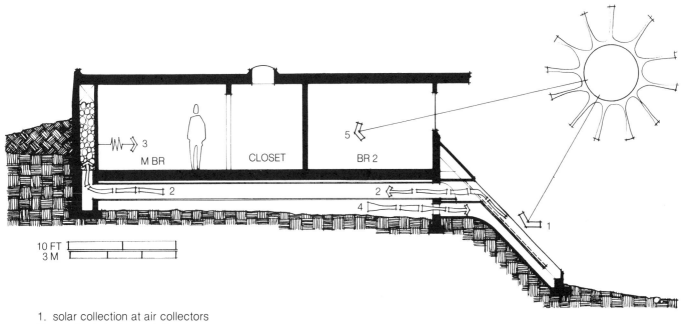

10 FT
3 M

1. solar collection at air collectors
2. supply air to thermal storage
3. radiant thermal storage mass
4. return air to collectors
5. direct gain to living space

Section

Project Data Summary

Project Information

Project: La Tierra residence
            Santa Fe, New Mexico
Architect: Mark M. Jones, AIA
            Santa Fe, New Mexico
Builder: Mark M. Jones & Partners
            Santa Fe, New Mexico

Climate Data

Latitude	36.0°N
Elevation	6,750 FT (2,057.5m)
Heating degree days	6,100
Cooling degree days	331
Annual percent possible sunshine	72%
January percent possible sunshine	62%
January mean minimum outdoor air temperature	16°F (−9°C)
January mean maximum outdoor air temperature	41°F (5°C)
July mean minimum outdoor air temperature	56°F (13.5°C)
July mean maximum outdoor air temperature	84°F (29°C)

Climate features: dry, high altitude

Building and System Data

Heated floor area	2,134 FT² (198m²)
Solar glazing area	
Thermosiphon air collectors	576 FT² (53.5m²)
Greenhouse	210 FT² (19.5m²)
Direct gain	42 FT² (4m²)
Thermal storage heat capacity	
Vertical rock bed	13,680 Btu/°F
Horizontal rock bed	3,360 Btu/°F
Adobe mass wall (greenhouse)	2,530 Btu/°F
Concrete slab and brick	11,230 Btu/°F

Performance Data

Building load factor	5.4 Btu/DAY°F FT²
Auxiliary energy (heating)	7.8 MMBtu/YR
Auxiliary energy (cooling)	2.9 MMBtu/YR
Solar heating fraction	87%
Night ventilation cooling fraction	93%

HOWIE RESIDENCE: Raleigh, North Carolina

Much publicity has been accorded the so-called envelope house. Howie residence, Raleigh, North Carolina.

The double envelope house has received a great deal of publicity and notoriety from the media. One person who has worked with this concept is architect Lee Porter Butler of San Francisco. He has organized Ekose'a, a national marketing firm which modifies its basic design for different climate zones. Regional representatives sell the designs and advise during construction.

Butler states, "The Ekose'a house has been described as a double exterior-shell 'house within a house.' In a conventional house, the walls, floors, and ceilings are exposed directly to the extremes of the outside climate. In this design, a convection envelope is created by separating the interior surfaces from the exterior surfaces with a protective blanket of moving tempered air. The interior walls and other surfaces are thereby isolated from the extremes of the outside climate.

"Since the ground temperature at the crawl space remains reasonably stable and is only a few degrees below the temperature required for human comfort, the heat generated by electric light and people is sufficient to maintain comfortable interior conditions much of the time. During the most extreme winter conditions, a fire in the fireplace or small heaters are sufficient to makeup the additional heat loss."

Some valid technical questions have been raised about the double envelope design. There are a number of prevalent misconceptions and claims that must be addressed and clarified before an understanding of the advantages and disadvantages of this concept can be verified.

1. How does the convective air flow within the envelope work?
2. Does adequate thermal storage occur in the crawl space?
3. Does the envelope space function to only limit, or does it distribute, building heat loss?
4. Is the added cost of construction justifiable in terms of performance and comfort?
5. Does this design require more or less energy to heat and cool than other passive or active solar designs?

There are also the following architectural features to consider:

1. The envelope should have a sufficient fire rating or controls to avoid rapid spread of flame or smoke around the structure.
2. Measures should be taken to prevent water absorption and buildup, which cause mildew within the envelope.
3. Shading of the greenhouse and, when required, ventilation of all spaces during warm weather are necessities to minimize overheating.

A number of qualified groups and individuals have investigated these and many other aspects of the double envelope concept. The conclusions are not cut and dried; it appears that the design can function effectively, but the concensus is that other passive systems can perform as well, if not better, depending on the design and the climate. There is no single best answer for solar space heating and cooling. Instead, there are probably as many design options as there are climates, styles, and budgets. How well suited these concepts are depends, in large part, on performance. Since passive solar architecture is a young and growing technology, many new ideas will be developed. The double envelope concept incorporated in this house is just one of the exciting possibilities, and, like any new concept, adds vitality to the maturing solar field.

From the Architect

John and Sistie Howie contacted me about designing a house for them on several acres in the southern pine forest outside of Raleigh. As part of my initial visit, I worked with them on selecting a site: a flat clearing in the woods with good solar exposure. The Howies wanted a fairly large house with a big kitchen and space for formal dining and entertaining. Along with a master bedroom suite, they wanted separate bedrooms for their two children, as well as a library and study.

The solution was a fairly straightforward, offset shed-roof design. It is compact in form and simple in detail, for reasons of both economy of construction and a low surface-to-volume ratio. It is two stories, with living areas on the first floor and bedrooms upstairs. Utility spaces, bathrooms, and closets were placed along the north as buffers, while more frequently used spaces face onto the solarium to the south. The original plan was to be a wood frame building with a crawl space below.

The design incorporated a full north and south envelope from the solarium through trusses in the roof, a double north wall, and the crawl space below. The site's hot and humid climate pro-

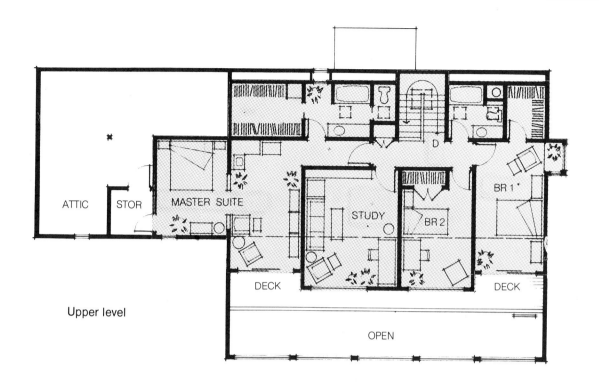

ATTIC   STOR   MASTER SUITE   STUDY   BR 2   BR 1

DECK   DECK

Upper level

OPEN

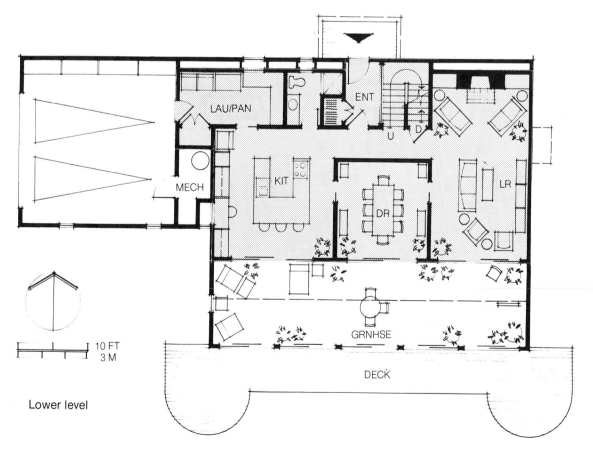

LAU/PAN   ENT

MECH   KIT   DR   LR

U   D

GRNHSE

10 FT
3 M

DECK

Lower level

vides a fairly severe test of my ideas on how the envelope circulation, underground cooling or make-up air, and ventilation function. The goal was to maintain interior comfort levels all year round in a totally natural, nonmechanical manner.

Performance has yielded mixed results. The coldest interior temperature the Howies have experienced is 60° F (15.5°C) on a morning with a 9° F (−12.5C) outside temperature, after a two-week overcast period. Supplementary heat for the home was to be provided by one of the latest prefab fireplace units, but it proved to be nothing but trouble. The outside

source of combustion air recommended by the manufacturer was apparently insufficient as the unit created drafts throughout the house when it was in use. Other clients have reported similar experiences, and we have some doubts about the usefulness of these units.

Summer performance has revealed some serious shortcomings. The cooling tubes have performed quite well, supplying mid-50° F (12–14°C) air to the base of the building at the start of summer and, as ground conditions changed, around 70° F (21°C) air toward the end of the season. The principal problem has

been with the ventilation, or, more accurately, moving the cool air through the building effectively. I think we put too much faith in the solarium ventilation; the upstairs living level has run quite hot all summer. The hottest conditions were during the day, fortunately, when these areas are not much used.

Humidity levels inside were also quite high; however, a foundation leak resulted in standing water in the basement, and it was not removed until halfway through the summer. Also, some of the trees immediately to the south of the building were removed during construction; these were critical to the summer shading, since I had designed the building without any external shading devices. This error obviously is the source of a lot of overheating, and some provisions will have to be taken to correct it. I do feel there are ventilation problems that must be overcome; if we cannot arrive at a simple architectural solution, a fan system may be the most practical answer.

From the Inhabitant

We have committed ourselves to building and living in a design of this type in order to understand and experience its performance. After living in the house for a full season, we are involved in modifying the design. Our goal is to make the house do what it is supposed to do—heat and cool itself naturally.

We have experienced some successes and some problems. The house is nice to live in during the winter. The temperature inside is unusually comfortable, and the interior humidity during winter is higher than in a conventional home: a benefit in this climate. Also, the greenhouse is a pleasant part-time living space, which we had not anticipated. The greenhouse temperature has gone as low as 48° F (9°C), but generally it is warm and sunny during the day. We

Space between the north windows aids envelope convection. Howie residence.

The greenhouse provides both primary solar collection and a living space.
Howie residence.

are hoping to adjust it to maintain temperatures above 68°F (20°C), which we prefer. Because the so-called high-efficiency fireplace doesn't heat the house well, we have used portable electric heaters when necessary and have accepted higher temperature swings. When the greenhouse temperature goes above 70°F (21°C) we open its doors to allow the heat to flow into the living area.

We made the mistake of installing a full basement under the house. This was supposed to have been a crawl space for the air to circulate on its way back to the bottom of the greenhouse. We have hung a polyethylene membrane under the floor joists, creating a

1. winter solar gain
2. warm convective air flow
3. convective flow below
4. return air flow
5. convective flow through living space
6. outside air intake
7. warm-air exhaust

12-inch (30.5cm) air space. This does not work well. The architect has told us that it is important for the air to have contact with the earth, and we may have to fill the basement in. In retrospect, we would not have built the basement, since the greenhouse provides the additional space we need.

The large greenhouse, with its forty-five-degree roof pitch, loses a lot of heat in the winter and acts like a big heater in the summer. The greenhouse has double exterior glazing and single glazing next to the living space. An interior awning is being designed to prevent heat gain to the house and overheating of the greenhouse. The quality of natural light to the interior of the house is great.

The air flow around and through the house doesn't always go where it's expected. This is the biggest single problem in the summer. Warm air from the

greenhouse is supposed to exit out the top vents and draw cool air from the underground cooling tubes through both the greenhouse and the living spaces. For some reason, the dominant flow is through the greenhouse and not through the living areas. It sometimes gets very hot upstairs. We are using a number of fans in the loft and greenhouse to help ventilate the house and sometimes open outside north windows to assist with cooling. A thermal chimney for the greenhouse is being designed, which we hope will replace the fans.

If we were to build again, we would make the house smaller, primarily by eliminating the basement. However, we are dedicated to making adjustments to the house. It is interesting to discover the effects modifications will make and to anticipate what the interior air currents will do.

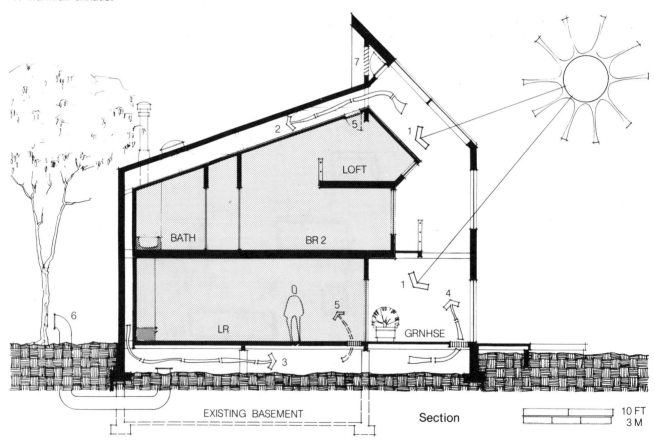

Project Data Summary

Project Information

Project: Howie residence
            Raleigh, North Carolina
Architect: Ekose'a—Lee Porter Butler
            San Francisco, California
Builder: Jet Barker with Sistie Howie
            Raleigh, North Carolina

Climate Data

Latitude	35.0 °N
Elevation	400 FT (122m)
Heating degree days	3,514
Cooling degree days	1,394
Annual percent possible sunshine	61%
January percent possible sunshine	56%
January mean minimum outdoor air temperature	31°F (−.5°C)
January mean maximum outdoor air temperature	52°F (11°C)
July mean minimum outdoor air temperature	68°F (20°C)
July mean maximum outdoor air temperature	88°F (31°C)
Climate features: moderate winters; hot, humid summers	

Building and System Data

Heated floor area	2,771 FT^2 (257.5m^2)
Solar glazing area	
Convective envelope, tilted glazing	500 FT^2 (46.5m^2)
Convective envelope, vertical glazing	490 FT^2 (45.5m^2)
Thermal storage heat capacity	
Concrete foundation	Unknown
Earth	Unknown

Chapter 5

SYSTEM COMBINATIONS

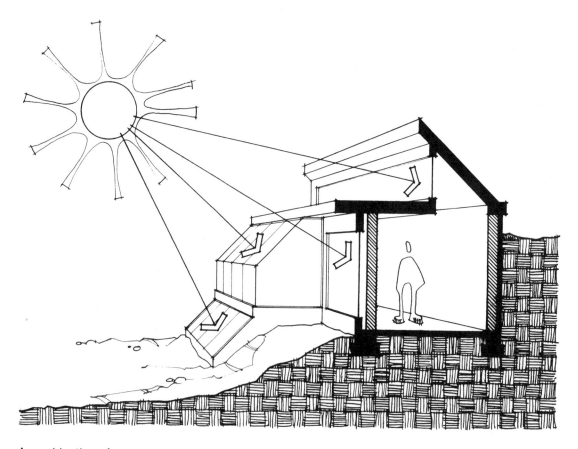

A combination of systems

Many passive solar buildings combine features of direct-, indirect-, and isolated-gain systems. Each feature is incorporated with varying degrees of emphasis and for particular design reasons. For example, a mass wall may be used in a living room to prevent daytime overheating or sunlight damage to furnishings, while direct gain may be desirable in a bedroom to capture the first rays of daylight for both heating and awakening.

Any combination of these general passive systems may stem from program requirements or designer preference. Each system type is selected primarily for its thermal and architectural effect. The appropriate selection and combination of these systems allow the logic and beauty of each building to emerge. The puzzle is complex and the pieces are many. The skill of the designer, who selects, proportions, and integrates these pieces, determines a design's success. And the greater the number of passive solar elements, the greater the challenge to combine them well.

Solar remodeling gives this house a new distinction. Pfister residence retrofit,
Minneapolis, Minnesota.

A variety of passive-solar techniques transformed this 1920 frame house into a model of energy-conscious living. Pfister residence retrofit.

It is estimated that more than 70 percent of all single-family residences that will exist in the year 2000 are already built. In order for passive-solar and energy-conscious design principles to have an impact on the residential sector, renovating and weatherizing of existing homes must occur. To date, however, little creative effort has been expended on this relatively untapped resource.

As the owner of a 1920 residence of traditional design, architect Peter Pfister set out to prove the passive retrofit potential in the cold Minneapolis climate. The first need was a tight weatherskin. Second was a change of focus of the house, from the street side on the north to the pleasant south backyard. This provided privacy, view, and the potential for significant passive solar heating.

Through weatherization, insulation, passive solar heating, and the addition of a little floor space, winter heating needs have been reduced by about two-thirds. Other gains include a strong visual connection of interior spaces to the private backyard and ample winter daylight, all vastly enhancing the comfort of this Minnesota renovation. The integration of a few "high-tech" devices, some contemporary design ideas, and a good feeling for tradition make this house a benchmark in the transformation of existing buildings into energy-conscious ones.

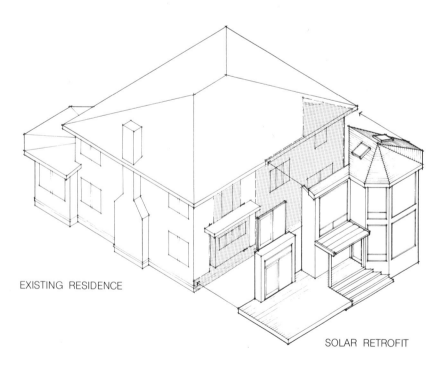

EXISTING RESIDENCE

SOLAR RETROFIT

Solar addition integrates with south facade of existing building.

From the Architect

My wife, Darlene, and I purchased a two-story wood-frame 1920 residence in south Minneapolis. It was essentially uninsulated. To begin the weatherization project, we insulated the 4-inch (10cm) frame wall cavities by blowing in polystyrene beads from the interior and then painting the southeast surfaces with vapor-barrier paint. The attic insulation was upgraded from 2 inches to 12 inches (5 to 30.5cm) of fiber-glass batting. Insulated shades were installed over several windows. This additional insulation, weatherstripping, caulking, and installation of water-saving shower heads and laundry fixtures reduced our 1978–1979 annual gas costs from $750 to $300.

The house is located on the south side of an east-west street, with the basement-level garage opening to the street on the north. The rear yard is particularly pleasant, but the house failed to take advantage of this view and privacy. The building had very few windows on the south, the major windows being oriented to the north and west. The major thrust was to increase the glass area to the south, increase thermal mass within the building, and improve the relationship between the living areas and the rear yard.

Other goals of the project were the following:

1. Design and install a simple passive system with typical components that can be used on other retrofit projects.
2. Optimize the use of passive solar gain.
3. Interface with the existing heating system.
4. Improve the function and appearance of the existing residence.
5. Improve cross-ventilation for summer cooling.

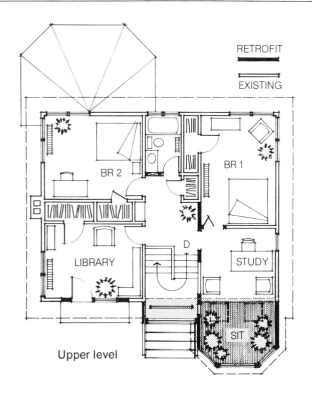

RETROFIT

EXISTING

Upper level

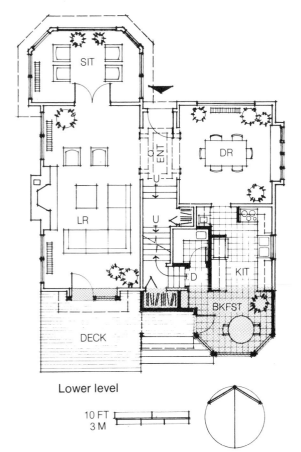

Lower level

10 FT
3 M

The retrofit consisted of the addition of two primary solar-gain areas and two secondary solar-gain areas. A 9-by-12-foot (2.5-by-3.5m) two-story solarium was added to the kitchen/breakfast area at the southeast corner. This sunspace allows direct gain to the building and allows the warm air to rise through a steel-grate second floor to the ceiling, where it can be circulated by a small fan to other spaces on the north side of the residence. The upper steel-grate floor also allows low-angle winter sunlight to filter deep into the kitchen space below. All windows in this space are equipped with motorized, roll-down insulating shades. The ceiling of the former garage under the kitchen was constructed of 10-inch-thick (25.5cm) concrete to serve as a fire protection. We insulated this concrete underneath and it is now used as thermal storage.

The second major passive-gain area was integrated into the stairway landing leading to the second floor by projecting a full-height window bay. The area behind this glass contains a phase-change thermal storage medium—twenty-five rods absorb radiant solar energy during the day to control overheating and release heat to the inside at night. This glazing is also insulated at night with motorized roll-down window insulation. External reflectors provide extra solar gain.

Two additional direct-gain areas were incorporated into the south wall of the living room and the bedroom above. A deck was added to the area off the living room and is reached from the living room and breakfast areas.

The existing heating system consisted of a gas-fired boiler with a gravity hydronic radiator distribution system. Since this system does not recirculate air, the distribution of heated air was a primary concern. The design solution allows warm air to rise through the metal-grate flooring to the top of the two-story solarium, where it

can circulate to the north side and interior portions of the house via cutouts in the adjacent walls. A small fan draws warm stratified air from the ceiling of the central stairway and returns it to the north side of the living room and dining room. Return air to complete the air loop is drawn by another small fan off the floor of the living room and past the thermal storage rods at the stairway bay.

From the Inhabitant

It didn't look like the house of our dreams. The living room was black and blue, the dining room green and brown, and every one of the fifty windows and doors was curtained. The blue and green metallic wallpaper in the stairway didn't do much for us either. But the rooms were large, the structure sound and well cared for. Most important, the house had a north-south orientation with room to expand.

First we insulated. Using a large cardboard box and a jerry-rigged vacuum cleaner, Pete devised a method of blowing polystyrene beads into the interior wall cavities through holes drilled between the studs. Because we intended to repaint every room, this was much easier to do from the inside of the house than through the stucco outside. And because we could do it ourselves, it was inexpensive—about $500 for the beads and the fiber-glass batting to insulate the attic.

Of course we were pleased by the energy savings due to these simple conservation measures, but an architect, especially one whose main interest is solar design, can hardly rest when a south-facing wall is not open to the sun. Ours wasn't. So we began tearing parts of the existing south wall down and then, with two carpenters, we began putting it back together. The original small windows on the south

side of the house were replaced by a two-story glass-enclosed solarium. Besides the six windows of the solarium, there were two window-door combinations and a window in the stairway landing.

The new design of the house increased the amount of south-facing glass from about 50 square feet (4.5m^2) to 220 square feet (20.5m^2). Besides brightening the house considerably, the windows allow the low winter sun in to heat the space. Vegetation and overhangs shield the rooms from summer overheating. Heat is stored by the quarry tiles in the kitchen floor (which absorb the sun's heat and radiate it into the room when the temperature drops) and by the rods behind the stairway landing's big window.

While adding all those windows had many desirable aspects, it also presented a problem at night and on cold, overcast days. The windows let the heat out. To prevent excessive heat loss, all the new windows were fitted with four-layered, motorized insulating shades that operate on a thermostat. The shades open automatically to let the sun in and close when the sun goes down to hold the heat in. Existing windows were fitted with fabric-covered insulation that can be removed on warm, sunny days.

We installed an air-circulation system to keep heat from staying on the south side of the house and rising to the top of the second floor. We used a thermostat-controlled fan with a duct. Openings in the walls of the upstairs solarium and the bedroom allow the air to flow through the rooms. And with the help of a few open windows on each side of the house, these openings help keep the house cool and breezy during the summer.

Finally, we save a lot on energy. The house is much brighter and more pleasant to be in. We still have many things to do, but it is already becoming the house of our dreams.

The two-story solarium off the kitchen becomes the breakfast nook. Pfister
residence retrofit.

1. indirect winter solar gain
2. phase-change thermal storage
3. radiant heat
4. convective heat from thermal storage
5. interior convection
6. direct winter solar gain
7. fan for interior convection
8. fan for thermal storage convection
9. insulated shades

The steel-grate floor above the breakfast nook allows circulation of warm air.
Pfister residence retrofit.

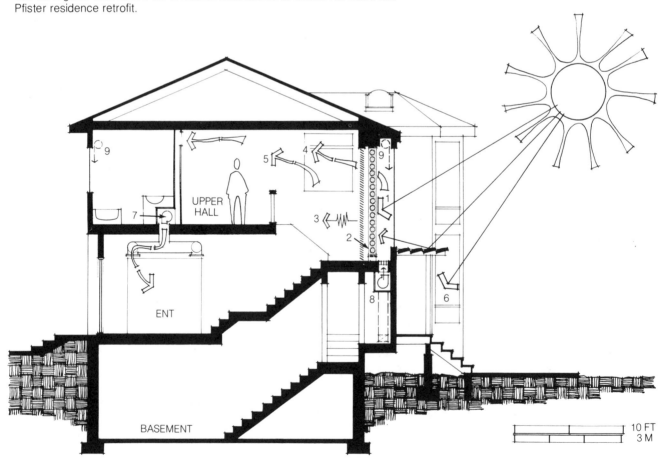

Section

Project Data Summary

Project Information

Project: Pfister residence retrofit
          Minneapolis, Minnesota
Architect: Architectural Alliance—Peter Pfister, AIA
          Minneapolis, Minnesota
Builder: The Pfisters and friends

Climate Data

Latitude	44.9°N
Elevation	255 FT (77.5m)
Heating degree days	8,159
Cooling degree days	585
Annual percent possible sunshine	58%
January percent possible sunshine	50%
January mean minimum outdoor air temperature	3°F (−16°C)
January mean maximum outdoor air temperature	21°F (−6°C)
July mean minimum outdoor air temperature	61°F (16°C)
July mean maximum outdoor air temperature	82°F (27.5°C)

Climate features: very cold, cloudy winters

Building and System Data

Original floor area	1,830 FT^2 (170m^2)
New floor area	200 FT^2 (18.5m^2)
Solar glazing area	
Direct gain	165 FT^2 (15.5m^2)
Phase-change rods	55 FT^2 (5m^2)
Thermal storage heat capacity	
Tiled concrete floor	1,600 Btu/°F
Phase-change rods	63,550 Btu

Performance Data

Building load factor	8.7 Btu/DAY°F FT^2
Auxiliary energy (heating)	46.6 MMBtu/YR
Auxiliary energy (cooling)	6.4 MMBtu/YR
Solar heating fraction	48%
Night ventilation cooling fraction	71%

The south elevation incorporates a seasonal greenhouse with direct gain.
Nourse residence, Worcester, Massachusetts.

The earth-integrated north elevation of a unique triangular solar home. Nourse residence.

A design which, through the use of shape, material, and layout, provides most of the space conditioning for a wide range of climate zones is laudable. A design called Trisol (patent pending), developed by Leandre Poisson, is a package intended to be sold at a modest cost.

This house is a direct-gain passive solar building which uses natural convection to assist thermal distribution and ventilation, augmented by a seasonal greenhouse. In winter, the easily assembled greenhouse acts as a sunspace attached to the south mass wall. In summer, it is removed to expose the uninsulated wall to the outside, helping to cool the kitchen. Like any passive solar house, however, the solar aperture, thermal mass, volume proportioning, insulation, shading devices, and ventilation openings must be refined to suit a particular location.

The building's fresh triangulated form is striking from the outside and defines unique living spaces. This design approach generates an architectural rethinking; it uses the principles of passive solar design, free from the restraints of construction standards and architectural clichés.

From the Designer

As a practicing professional, I was trained in twentieth-century design philosophies. Now that we find ourselves in the energy crunch, it has become obvious that some of these philosophies have contributed to the worsening crisis. The pseudoscientific and arbitrary esthetic postures that our solutions have been rooted in have resulted in dinosaur architecture—beasts with limited uses that consume enormous amounts of finite resources in their relatively short lifespans.

With this in mind, I set about to discover what buildings were in concert with the biosphere and also had a long history of utility. The answer was simply "vernacular or indigenous" architecture. These structures were evolved by people who were rooted to a locale and wedded to its finite resources. Knowing this, I have limited my efforts to developing designs for New England that best meet the needs of living here—solutions that take into consideration local climate, traditions, materials, and building techniques, as well as our current scientific knowledge.

The Nourse house was the first result of this effort. Trisol was conceived primarily as a passive "solar device" that would meet the solar needs of the inhabitants with minimal input of outside resources and energies. The secondary consideration was living space. The solar needs are defined as food production and preservation, space heating and cooling, cooking, and domestic water heating.

The house is a device in the full sense of the word: that is, the shape, profile, and configuration directly contribute to its solar function. For example, the triangular plan of the house was arrived at because it offers the best ratio of collector to storage mass to living space, not for esthetic reasons.

The highest point in Trisol is at the rear of the house. This configuration corrects what I believe is an error found in much solar thinking: a high front glass area with the roof sloping down to the lower rear wall. This profile results in the warmest air sitting up against the poorly insulated glass, losing heat to the outside day and night. In my design, the warmest air is conveyed to the back of the house, away from the windows. Because of the triangular plan and the pitch of the ceiling, the convective force of the rising air drives warm air down the back channel. No fans are needed. In the summer, the warmest air is ventilated at the high point, forcing in precooled air through earth-covered intake pipes, open to the outside.

The greenhouse optimizes the combined functions of growing food, heating space, and heating water. It can produce sufficient salad vegetables even during the winter. Later designs include a walk-in root cellar accessible from the kitchen and located in the side earth berm. A wood-burning stove is utilized for food preparation and also serves as the only backup heating source.

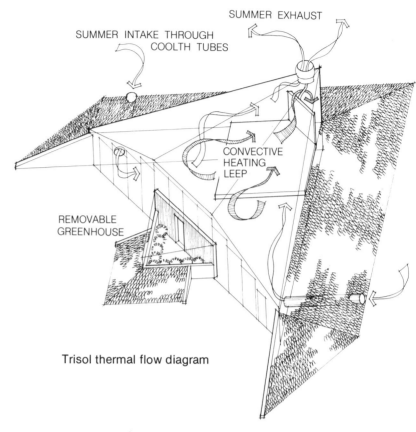

SUMMER EXHAUST

SUMMER INTAKE THROUGH COOLTH TUBES

CONVECTIVE HEATING LEEP

REMOVABLE GREENHOUSE

Trisol thermal flow diagram

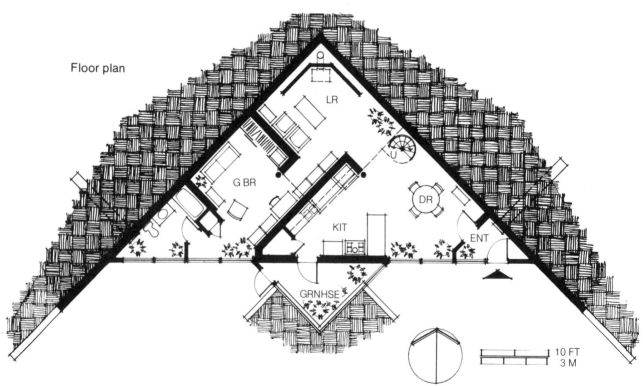

Floor plan

LR

G BR

U

DR

KIT

ENT

GRNHSE

10 FT
3 M

From the Inhabitants

Our home is a very unique place to live. Although it is an average size—about 1,400 square feet (130m²)—the openness and tremendous amount of light give a grand feeling of spaciousness and flow. Each season, as well as each time of day, offers a varying feeling of the interior space. Part of this is due to the building's openness to the outdoors. Nature is virtually our wallpaper. The entire interior focus is to the south wall. In summer, we can gaze at a vegetable garden, strawberry and hay fields, or at flower beds from every point within the house. The fall foliage and winter snows offer a variety of outdoor textures and perspectives. In all seasons, the sky is visible from each section of the house.

Day and night, morning and evening vary with the casting of the sun's or moon's rays directly into the house. In all seasons, our house operates well. On an extremely cold sunless winter's day or evening, we need a small fire in our wood stove to keep the temperature near or above 70°F (21°C). In summer, we find the temperature inside the house is always lower than that outside.

Our home costs very little to operate. Our highest monthly utility bill has been $35 for electric lights, appliances, and hot water. We expect this figure to be reduced by at least one-third once we install our solar water-heating system on the greenhouse roof.

The highest point of the roof is at the rear, to convey warm air. Nourse residence.

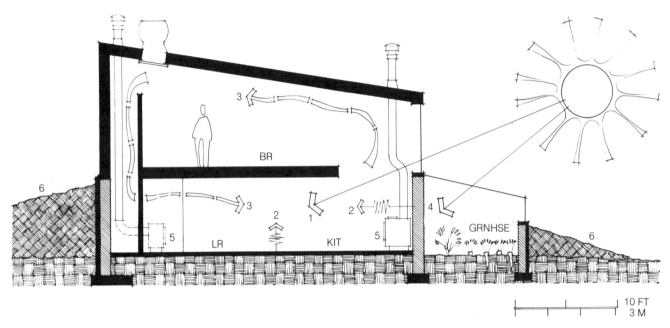

Heating mode

1. direct winter solar gain
2. radiant thermal storage mass
3. convective air flow
4. indirect winter solar gain
5. energy-efficient wood stove
6. earth integration

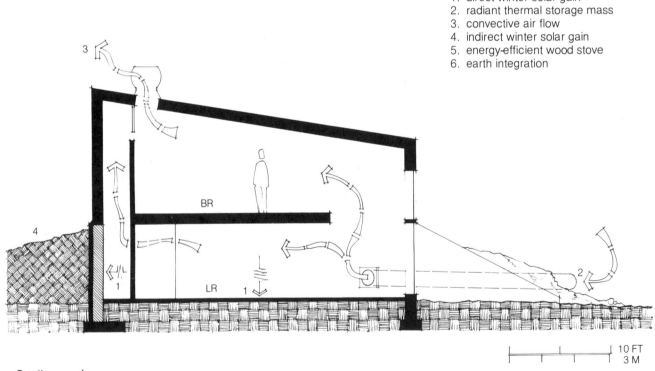

Cooling mode

1. cool mass absorbs daytime heat
2. outside air intake
3. warm-air exhaust
4. earth integration

Project Data Summary

Project Information

    Project: Nourse residence
            Worcester, Massachusetts
    Designer: Trisol—Leandre Poisson
            Harrisville, New Hampshire
    Builder: John and Malsan Nourse
            Worcester, Massachusetts

Climate Data

Latitude	42.3°N
Elevation	300 FT (91.5m)
Heating degree days	7,400
Cooling degree days	251
Annual percent possible sunshine	55%
January percent possible sunshine	45%
January mean minimum outdoor air temperature	17°F (−8.5°C)
January mean maximum outdoor air temperature	31°F (−.5°C)
July mean minimum outdoor air temperature	61°F (16°C)
July mean maximum outdoor air temperature	79°F (26°C)
Climate features: harsh winters; mild summers	

Building and System Data

Heated floor area	1,418 FT² (131.5m²)
Solar glazing area	
Direct gain, greenhouse	600 FT² (55.5m²)
Thermal storage heat capacity	
Tiled concrete floor	10,890 Btu/°F
Interior and exterior masonry walls	26,510 Btu/°F

Tucked into a hillside on the Aspen mesa, this house is elegantly energy conscious. Mulford residence, El Jebel, Colorado.

The Rocky Mountains challenge passive solar architecture with bitterly cold winters. But the region is also blessed with above-average winter sunshine, and at an altitude of 7,500 feet, you are closer to the sun.

The Mulford residence is tucked into a hillside on Colorado's Aspen mesa, echoing the rock outcroppings of the surrounding mountains. From the north, the house is subterranean, with sod roofs to protect it from icy blizzards and to blend with the environment. Built by three masons, the exterior walls are exposed jumbo brick, which provides the finished interior wall surface. These masonry walls are insulated on the outside with three-inch rigid insulation sheathed with wood above grade.

Architect Peter Dolbrovolny has tastefully combined three passive solar systems to form the south facade, which steps rhythmically along the hillside. Trombe walls, direct-gain glazing facing both east and south, and a greenhouse combine to collect and supply heat to the structure. The building is oriented fifteen degrees west of south to increase the amount of sunshine into the east windows and reduce afternoon shading by the stepped south wall.

Solar collection surfaces need not be colored black to maximize absorption; many playful color treatments are possible. Here, the Trombe walls are painted dark blue on the exterior to harmonize with the sky and mountain reflections off the glazing. The open-cell structural block walls of the lower-level bedrooms are linked to a ceiling plenum, which allows warm air to be drawn through it from the south Trombe wall to heat both rooms by radiation. Most mass walls are penetrated by south window openings to allow for daylight and viewing, some with masonry arches decorated with stained glass. The Trombe walls are automatically insulated at night by self-inflating curtains consisting of several layers of Mylar® film. When

the curtains are lowered air naturally rises between each layer, fluffing them apart, an insulative barrier of several air spaces is created; when the sun provides enough heat to warm the thermal mass, an exterior sensor activates a motor to raise the curtains. They close automatically when solar radiation is insufficient for heat collection.

Excess heat from the greenhouse is blown by a fan to a vertical gravel-filled wall on the north side of the greenhouse's masonry wall. Since the gravel inside the wall is insulated from the greenhouse, this stored heat is radiated directly to the north side of the house. Additional heat is provided by a fireplace inspired by the historic Count Rumford fireplace. This utilizes a shallow sloped-back firebox, a tapered brick-lined chimney, and an outside air source to increase efficiency. Water is heated by an active solar water-heating system. The solar collectors are the only tilted element

The sheltered northwest entry makes the home appear almost subterranean. Mulford residence.

The greenhouse is warm and sunny even on cold winter days. Mulford residence.

on the south facade, sloped for optimum year-round performance.

The striking interior finish of the passive solar elements creates a building that performs well and has surprise and warmth. During the building's first winter, the electric baseboard auxiliary heater bill was $2 in December and $11 in January—not bad for one of the severest winters on record.

From the Architect

Our clients wanted to construct a dwelling that went beyond standard contemporary design. A house that was distinctive, even arrogant, and decidedly energy-conscious; a house that would appeal to the affluent tastes of the "Aspen crowd." Ron Shore, a pioneer in passive solar components, and I spent three days on the site with drawing boards, conceiving the preliminary design. Lack of a definitive program and the builders' desire for a creative statement allowed us to make performance and passive solar expression the design criteria.

Familiarity with France's Trombe wall residences, designed by Jacques Michel and Felix Trombe, prompted us to modulate the usually monolithic south facade of such buildings with curved and forty-five-degree-angled walls. Previous to this project, I had not sufficiently mastered the tools of passive design to make that logical and simple step, which has led to greater freedom of architectural expression.

The Mulford residence, along with other passive projects, provides me the continuing personal gratification that I think may not be experienced by many architects not dedicated to this kind of work. My reward comes most importantly from the shared joy of clients and the pleasant surprise of introducing passive solar designs to the contractors with whom I have worked.

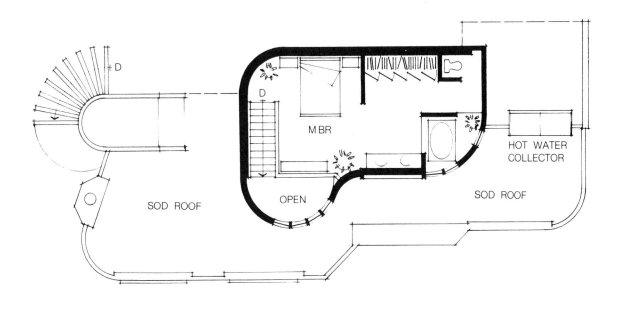

Upper level

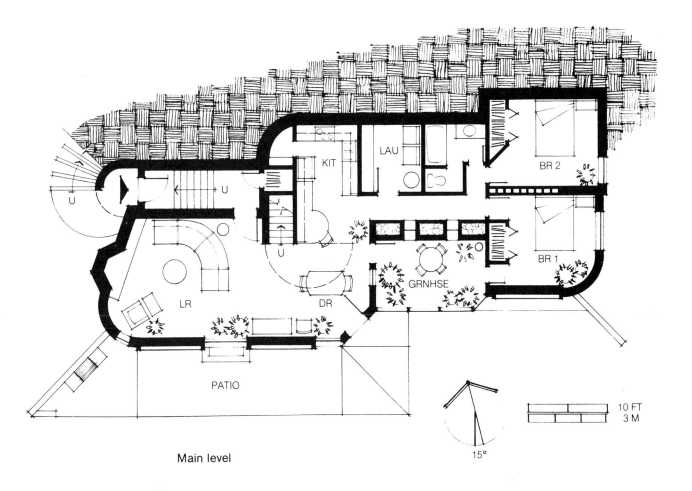

Main level

This design has outperformed our expectations of comfort and thermal performance. The winter of 1978–1979 was not kind the Colorado chamber of commerce likes to advertise: January was unusually cloudy, and the winter was the coldest in fifty years, with record snow. The house's performance demonstrated the viability of its design approach in less than sunny climates. Comparisons with my own house, which is both passive and active, clearly demonstrate the significant advantage that passive systems have over active in collecting and storing energy even on cloudy days.

Because it was a speculative residence and toured by many people, we heard varying opinions. The richness of materials and architectural shape caused many "wow" reactions and comments about how it was "just like a Hollywood house." Some complained that the Trombe walls obstructed views of the mountains to the south. Probably the most amusing and personally frustrating remark came from a local banker, who said, "I wouldn't live in one of those funny-looking solar houses even if the government paid me to."

We have successfully incorporated many of the passive elements into conventional construction means using standard materials. We currently have several projects built at costs equal to conventional housing, all having thermal performance comparable to the Mulford residence. Regardless of this record, the transition to solar housing is painfully and frustratingly slow. Many people, even in a small valley such as this (which had 117 solar residences by 1979), still don't understand or won't accept that passive solar houses really work.

1. indirect winter solar gain
2. radiant heat from Trombe wall
3. convective air flow
4. automatic insulating
5. direct winter gain to greenhouse
6. radiant thermal storage mass
7. forced air to rockbed
8. fan
9. return air to greenhouse
10. radiant heat from rockbed
11. earth integration

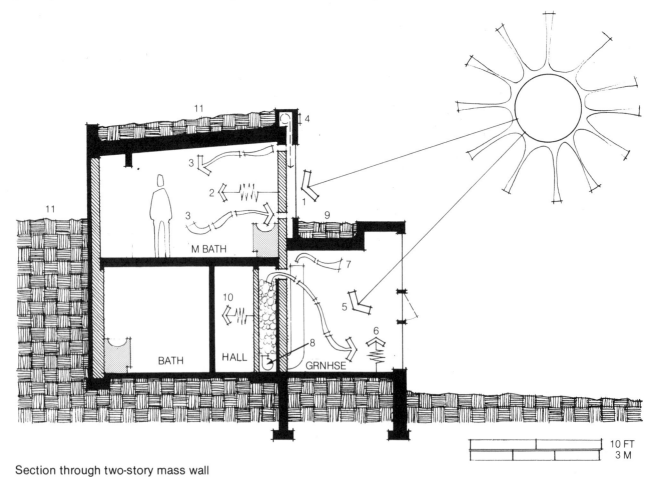

Section through two-story mass wall

10 FT
3 M

1. indirect winter solar gain
2. radiant heat from front mass wall
3. convective air loop
4. radiant heat from internal mass wall
5. automatic insulating shades
6. direct winter solar gain through clerestory
7. radiant thermal storage mass
8. earth integration
9. solar water heater collectors

The low curved wall over the dining area allows heat to be conveyed to master bedroom above. Masonry finish accents interior decor. Mulford residence.

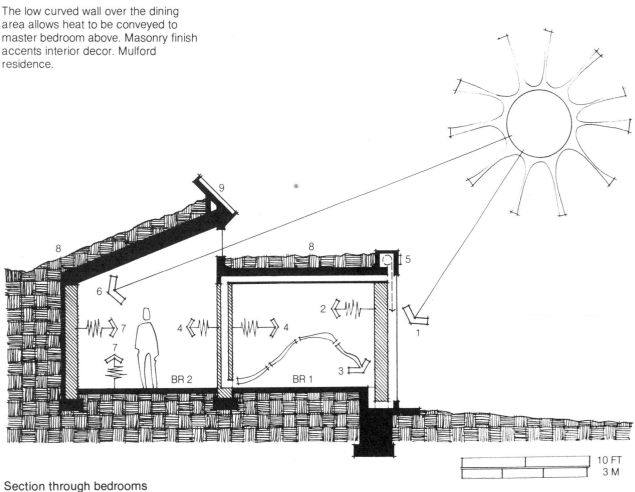

Section through bedrooms

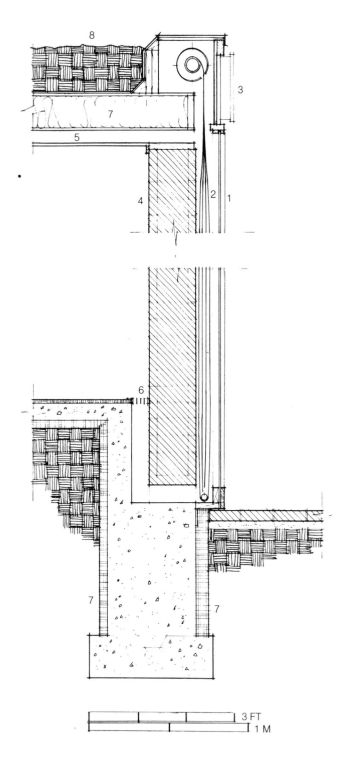

Modified detail of Trombe wall

1. transparent glazing
2. automatic insulating shade
3. shade-control sensor
4. mass wall
5. warm-air supply
6. air return
7. insulation
8. sod roof

From the Inhabitant

Passive solar design provides almost all space heating. Although I have not kept detailed records, we use the electric backup heat for only brief intervals during the winter. The passive solar systems provide comfort through the cold season, well into the third day without sunshine. Although I was accustomed to setting thermostats in the 68–70°F (20–21°C) range, this house is comfortable even down to 64°F (17.5°C). I think the masonry walls and floors radiating heat, together with superior insulation, caulking, and weather stripping, make the house comfortable at lower temperatures.

The house is relatively trouble-free. It has four Trombe walls with insulating curtains that come down automatically during sunless periods. After one season, we had pretty well mastered how to set the curtain's cycle and eliminated overheating in October and November, which we had experienced the first year. The house has a sod roof about 10 inches (25.5cm) deep, and I would not incorporate this feature in another home. The soil is not deep enough to retain moisture for extended periods and there have been leaks; repair or replacement is a horrendous job.

The Rumford fireplace works very well, but may need glass doors, like most efficient fireplaces. The shallow back reflects a great deal of heat into the room, but after you have retired for the night and the fire dies down, the hot air in the house escapes up the chimney and cold air comes down it.

Altogether, it is a delightful house to live in—very comfortable and easy to clean and maintain.

Project Data Summary

Project Information

Project: Mulford residence
        El Jebel, Colorado
Architect: Sunup Ltd.—Peter Dobrovolny, AIA
Builders: Jed Kairath, Thomas Gray, Ron Hoffman

Climate Data

Latitude	39.0 °N
Elevation	7,500 FT (2286m)
Heating degree days	7,340
Cooling degree days	0
Annual percent possible sunshine	70%
January percent possible sunshine	59%
January mean minimum outdoor air temperature	6°F (−14.5°C)
January mean maximum outdoor air temperature	36°F (2°C)
July mean minimum outdoor air temperature	46°F (7.5°C)
July mean maximum outdoor air temperature	80°F (26.5°C)
Climate features: heavy snow	

Building and System Data

Heated floor area	1,590 FT^2 (147.5m^2)
Solar glazing area	
Direct gain	165 FT^2 (15.5m^2)
Trombe wall	294 FT^2 (27.5m^2)
Greenhouse	155 FT^2 (14.5m^2)
Thermal storage heat capacity	
Trombe wall	7,450 Btu/°F
Concrete masonry, brick walls	41,310 Btu/°F
Tiled concrete floor	11,180 Btu/°F
Water storage	3,212 Btu/°F
Rock storage	2,220 Btu/°F

Solar art: a complex, ''holistic'' environment that makes use of the latest technology and esthetic. Crowther residence/research facility, Denver, Colorado.

Functional elements become architectural sculpture. Crowther residence/research facility.

Architect Richard Crowther has been designing energy-efficient structures for years. Like others, he considers the scale of residential buildings appropriate for studying and verifying concepts. He lives in a complex test facility, a proving ground for his ideas. All of Crowther's buildings are distinct and unique, and his residence/research facility is no exception.

Few elements stand independent of one another. Their interrelationship appears quite complex, yet the strength of the design is in this complex synthesis: Crowther has identified more than sixty solar concepts and nine basic subsystems in this building. Crowther has looked at the problems of skylights from an energy-waste standpoint and has designed and is evaluating his own "solar skyshaft," which insulates better than conventional skylights. Reflectors are used to melt snow in front of the garage, and one is placed on the roof drain to encourage water runoff. The water is collected in a 700-gallon (2,650l) cistern and used to water the landscape. Concerned about the negative bodily effects that electromagnetic fields may have, he has added wire mesh to the inner west walls and portions of the ceiling to ground them. The list goes on and on.

The building is essentially pre-stressed concrete with steel column and beam construction, plus a wood-frame clerestory and south wall. Because of the significant thermal mass and effectively placed insulation, Crowther is able to maintain thermal comfort with only minor interior temperature variations. Even though the total glazed area is only 12 percent of the total floor area, the annual solar heating contribution is estimated at 80 percent.

As creatively informal as the building may appear on the surface, there is a logic and order to the design. The spatial organization makes effective use of buffers to protect the interior spaces from the exterior. The rooms behave like ducts to channel air. The siting on this relatively small lot allows for privacy and good solar exposure on both building levels. And vegetation acts as a buffer.

Crowther combines not only the basic generic types of passive solar into his palette of architectural design, but also adds new ideas about heating, cooling, ventilation, and lighting. His complex interfacing of systems and their effects points the way to a more sculptural environmental architecture. As a live-in research facility, this building could well become a stepping stone for practical, energy-conscious, holistic applications in years to come.

From the Architect/Inhabitant

This residential passive and active energy research facility is located in an older urban neighborhood of Denver, Colorado. The project presently embodies a residence and a caretaker's apartment as well as a library, home office, solar test facilities, greenhouses, and indoor and outdoor recreational amenities.

The precast, prestressed concrete sections are the thermal mass. The residence is earth-sheltered at its north side as well as partially sheltered on the west and east elevations. South-facing elevations and a clerestory in the central gallery are wood frame. Exposed external walls are polystyrene insulation applied over the basic building envelope with a stucco-type finish. The roof is a reflective, aluminized material applied over polystyrene and wood fiber board. Concrete walls below grade are insulated with polystyrene board. The roof consists of a reflective deck and overhead angular reflective canopy. In winter, direct and reflected sunlight tends to optimize the solar impact on the fin-tube solar collectors and the concentration of solar radiation through the south clerestory windows.

The building, together with the topographic landscaping, internal occupant activities, and interior furnishings, forms a dynamic, interactive energy system. Climatic and earth-energy responses are calculated to seasonal change. The rapid swings of this Colorado mountain climate are attenuated by the design resolutions of building and site. As a research facility, the building has nine passive solar heating subsystems and one active liquid-type solar system. All systems can be monitored from a central control panel, which has a demand limiter to flatten utility-curve requirements and avoid peak electrical loadings. Each passive solar subsystem can perform independently or synergistically with other subsystems.

Properly proportioned glazed areas reduce energy losses at night and on overcast days, while transmitting ample solar gain to each solar subsystem. Although all window openings were designed for movable insulation, few have been provided with it. It is my opinion that all forms of movable insulation are relatively impractical. Who will bother to move it, who will be around to move it, how tight can you get it to fit, and how much insulation does it economically provide? Mechanical movement for such insulative applications is usually too costly, sometimes difficult to override, and requires maintenance and operational energy. Better solutions are needed, such as the high-transmis-

Roof monitor incorporates reflector, clerestory, and solar collectors. Crowther residence/research facility.

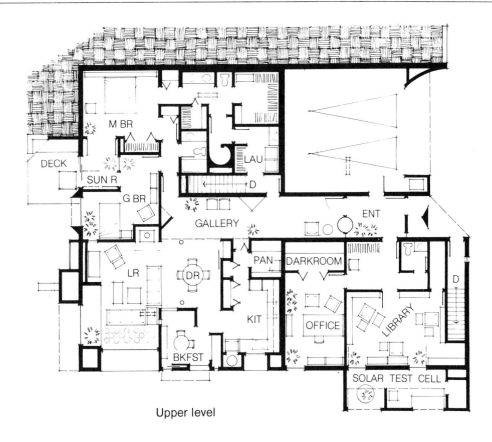

Upper level

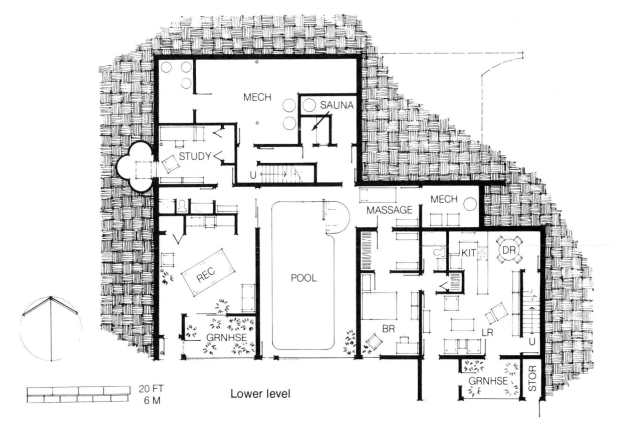

20 FT
6 M

Lower level

sivity films used in my southeast greenhouse glazing.

The upper-level residence area is designed to be heated and cooled by an auxiliary water-to-air reverse-cycle heat pump. Electric hydronic baseboard heating is a research alternative to the heat pump for the upper-level residence area. Total supplemental energy for heating 6,800 square feet (631.5m^2) of interior space, including the swimming-pool heater, is less than two watts per square foot.

Illumination of the interior space is augmented by the solar skylight (upper right). Crowther residence/research facility.

1. winter solar gain
2. radiant thermal storage mass
3. air-circulation ducts
4. solar collectors
5. solar storage tanks
6. solar reflectors
7. exhaust vents
8. skylight (solar skyshaft)
9. future sod roof
10. earth integration

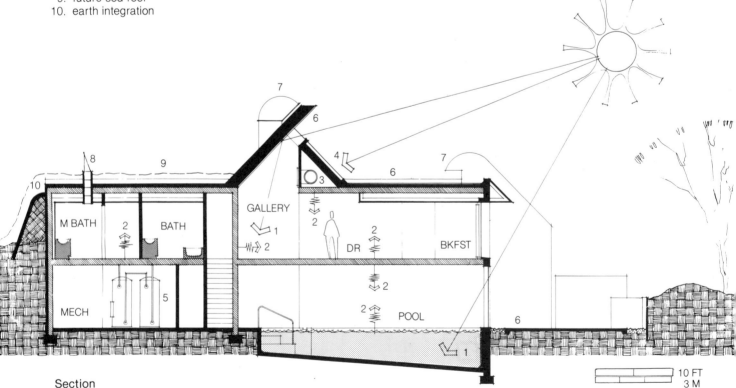

Section

The Subsystems

The following discussion is intended to give an idea of the complexity and design of Crowther's subsystems.

Subsystem 1. The lower southwest greenhouse has vertical acrylic glazing that receives direct and indirect solar radiation. Heat energy from the sun is stored in interior phase-change thermal storage tubes which reduce extremes of temperature fluctuation.

The white marble chips that pave both the outdoor courtyard and indoor greenhouse increase winter solar gain by reflectance. Solar radiation, by penetration through the inner glass panels that separate the greenhouse and recreation room, provides solar gain to the interior.

A flat black greenhouse return air duct from the upper level breakfast nook receives direct reflected solar radiation that, with phase-change thermal storage, can heat the upper level living room by forced air through

1. inductive solar chimney
2. solar skyshaft
3. roof vent turbine
4. exhaust air vent
5. water-collection basin
6. snowmelt reflector
7. subsystem 1
8. subsystem 2
9. subsystem 3
10. subsystem 4
11. subsystem 5
12. subsystem 6
13. subsystem 7
14. subsystem 8

Facility isometric

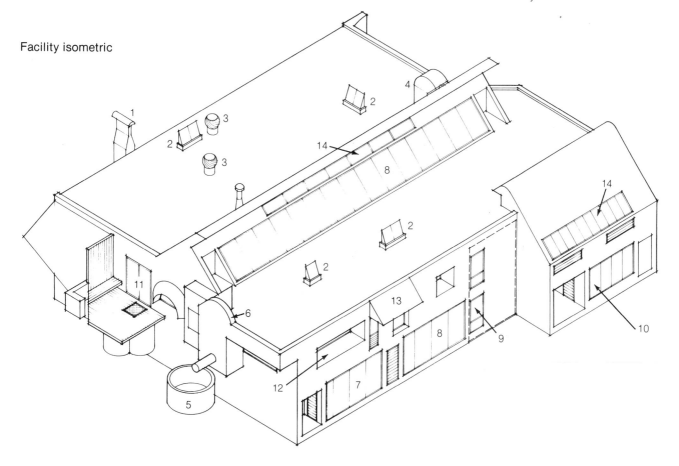

a high-efficiency filter. Humidity levels to this greenhouse can be controlled by removing the plugs to the air shafts that extend into the swimming pool area. Ventilation is achieved by a screened louver backed by an insulated metal door with magnetic weatherstripping. The inductive system includes a ductwork connection to a large roof downspout, seasonally operated with a manual damper.

Subsystem 2. The pool receives direct passive solar gain in winter through translucent solar glazing. The dark tile pool deck (over concrete slab) receives direct and reflected solar radiation from the white marble chip courtyard and interior white reflective surfaces. A liquid-type solar collector is mounted on the 45-degree angle roof. Its winter solar collection is increased by reflectance from the overhead canopy and roof deck. Solar heat is stored in a 1,100-gallon (4,164 l), underground water tank. From there, heat exchangers are used to heat the spa, swimming pool,

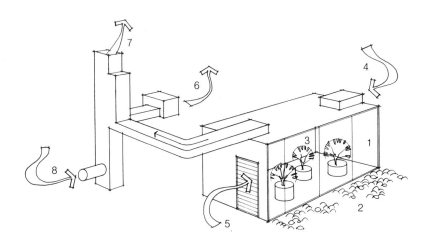

Subsystem 1: southwest greenhouse

1. transparent glazing
2. white marble chips
3. lunar glazing
4. greenhouse return air
5. ventilation
6. to upper level
7. exhaust air
8. air intake

Subsystem 2: indoor pool

1. translucent glazing
2. solar collectors
3. solar reflector
4. storage tank
5. heat exchangers
6. hot pool
7. swimming pool
8. air intake
9. exhaust air

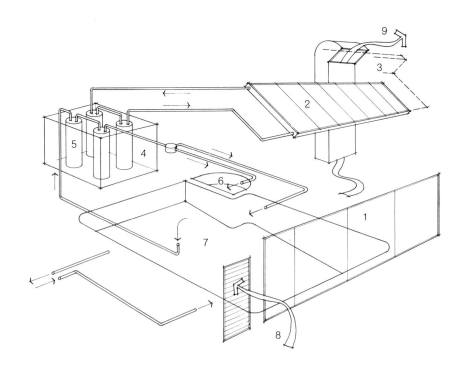

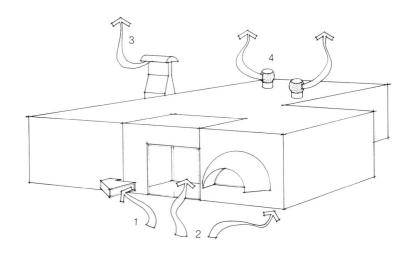

Subsystem 5: sunroom

1. screened intake air
2. outside ventilation air
3. inductive solar chimney
4. roof vent turbines

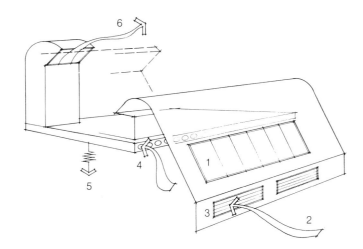

Subsystem 8: test deck (heating mode)

1. insulated glazing
2. air intake
3. automatic vents
4. air flow through concrete slab
5. radiant heat from thermal storage
6. exhaust air

Subsystem 8: test deck (cooling mode)

1. air intake
2. automatic vents
3. exhaust air
4. negative-pressure zone

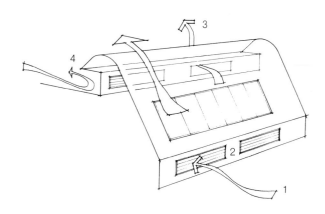

bathrooms, and domestic hot water. The swimming pool, located in the lower central area of the house, acts as a reservoir of ambient thermal energy and helps to heat the living space above.

Subsystem 5. This small room acts as a climatic buffer for the master bedroom and the guest bedroom. Direct solar gain is moderated by an interior translucent blind. A screened vent port is provided for the introduction of outside air to the guest bedroom. The master bedroom has an inductive solar chimney and an intake air vent with insulative panel.

Subsystem 8. Insulative sealed glass panes set at a 45-degree angle allow direct solar penetration during all periods of the year. The concrete thermal mass and dark painted wall receive direct solar gain during the winter months. Natural convection distributes heated air through the cores of the concrete ceiling, creating a radiative surface. For cooling, low solar vents automatically open to allow outside air to enter. The floor of the test deck is designed for experimentation with thermal storage and light reflective materials.

Project Data Summary

Project Information

Project: Crowther residence/research facility
Denver, Colorado
Architect/designer: Crowther Architects Group—Richard Crowther, AIA
Denver, Colorado
Contractor: Richard Crowther

Climate Data

Latitude	39.8 °N
Elevation	5,300 FT (1,615.5m)
Heating degree days	5,505
Cooling degree days	742
Annual percent possible sunshine	70%
January percent possible sunshine	73%
January mean minimum outdoor air temperature	15°F (−9.5°C)
January mean maximum outdoor air temperature	42°F (5.5°C)
July mean minimum outdoor air temperature	57°F (14°C)
July mean maximum outdoor air temperature	88°F (31°C)
Climate features: cold winters; dry, mild summers	

Building and System Data

Heated floor area	6,800 FT² (631.5m²)
Solar glazing area	
Direct gain	264 FT² (24.5m²)
Swimming pool	122 FT² (11.5m²)
Fluid solar collectors	360 FT² (33.5m²)
Thermal storage heat capacity	
Concrete mass	144,520 Btu/°F
Swimming pool	91,770 Btu/°F
Solar thermal tank	9,180 Btu/°F
Phase-change material (latent heat)	16,200 Btu

DEHKAN RESIDENCE: Andover, New Jersey

Not trying to blend with nature, this house is an unabashed solar statement.
Dehkan residence. Andover, New Jersey.

Many designers try to blend their designs with nature. On the other hand, buildings that contrast with their natural setting can produce more striking architectural statements with high thermal performance. This is man designing for man without imitating nature. As architect Doug Kelbaugh describes the Dehkan residence, "The building is unabashedly a man-made object abruptly inserted into the natural landscape." The building makes no attempt to soften, integrate, or metabolize with the environment; there is no sign of the organic romanticism often associated with passive solar designs. Yet it is an obvious success.

Kelbaugh looked at houses built in the area since World War II and attempted to provide a solution catering to the tastes and budget of most local home owners. The house was designed as a passive solar prototype for its climate, suited for a family with one or two children. It is a simple plan with simple zoning—living on the west, activity spaces and the sunspace in the center, and private zones to the east. The simplicity of plan and form accounts for the building's high thermal performance as well as for the owner's delight.

The solarium integrates water storage, venting, and shading. Dehkan residence.

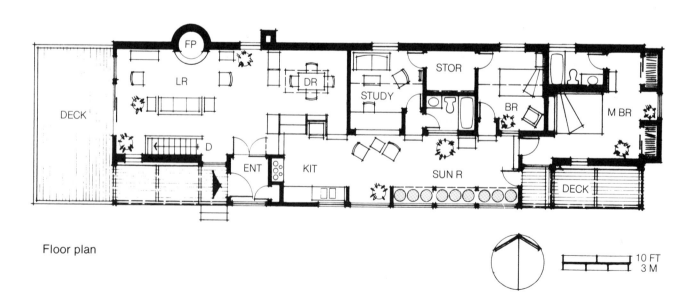

Floor plan

Kelbaugh is noted for his Trombe wall designs, and this project is no exception. However, it goes well beyond a simple single system approach; combined with the Trombe walls are drum walls, roof monitors, clerestories, and a solarium. Aiding these primary elements are large and carefully placed reflective surfaces to enhance solar gain. The preheat tank, suspended under the double-acrylic skylight dome, is coated with a selective surface to maximize net collection and reduce long-wave reradiation to the outside, thus improving efficiency. Selective surfaces, phase-change thermal storage materials, and other new products are being applied to many passive solar buildings to further simplify and improve their performances. The Dehkan house bluntly applies varied systems with special materials in an energy-conscious design for a particular program, site, and climate.

From the Architect

One design goal in the Dehkan house was to achieve a low-profile, horizontal house that is clearly inserted into the top of a small hill. With good views to the south, east, and west, it was important to part the Trombe walls and slide them to either end of the house, opening up the center for direct gain and a view. This central solarium was meant to be the hub of domestic activity—a circulation space, kitchen, greenhouse, and eating area. By raising the rooms which surround it a couple of steps, the solarium became an area into which six tributary spaces flow.

Because this was intended to be a high-performance house, we originally designed a vertical rock bed along much of the north wall. Excess solar gain was to be recaptured off the peak of the roof monitors, with the warm air blown down through the rocks to reenter the rooms at floor level. The north wall would then slowly release the heat stored within the rocks behind it. Although this actively charged, passively discharged rock bed was a sound design idea, it was dropped because of budget restrictions. We felt that there was enough thermal mass in the floor slabs, Trombe walls, mass walls, and water drums.

The owner and his carpenter came up with a clever sliding grate by which

1. winter solar gain
2. radiant thermal storage mass
3. natural convection
4. insulated shutter and reflector
5. sun-control blinds
6. operable vents
7. solar reflectors
8. earth integration

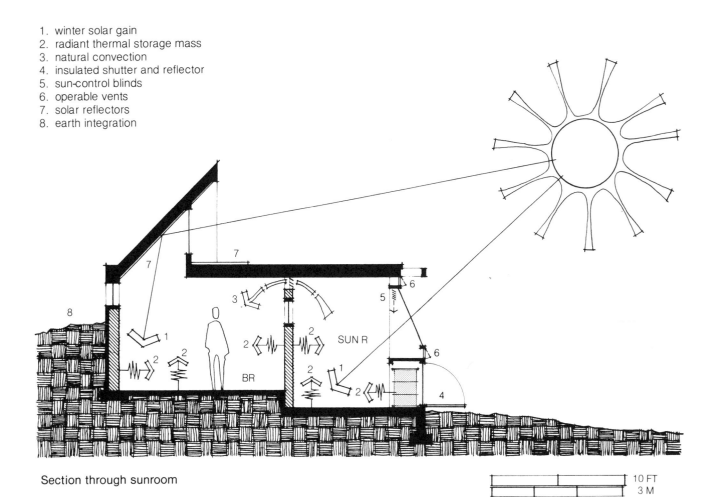

Section through sunroom

10 FT
3 M

one can close off the upper vents and effectively convert the system to a "stagnating" or unvented Trombe wall (solar mass wall). This strategy prevents daytime overheating on unseasonally warm, clear days and allows more heat to be stored in the wall for use at night.

From the Inhabitant

The house is a delight. We have become very aware of what is going on outside. The first few days we lived in the house, before there were any interior shades, the outside temperature went down to 20°F (−6.5°C). Inside, the house was so warm we decided to go ice skating to cool off! Anyone in New Jersey would be happy to have that kind of complaint in winter.

The unvented mass wall prevents daytime overheating. Dehkan residence.

1. indirect winter solar gain
2. radiant heat from Trombe wall
3. natural convection
4. insulated shutter and reflector
5. direct winter solar gain
6. solar reflectors
7. radiant thermal storage mass
8. earth integration

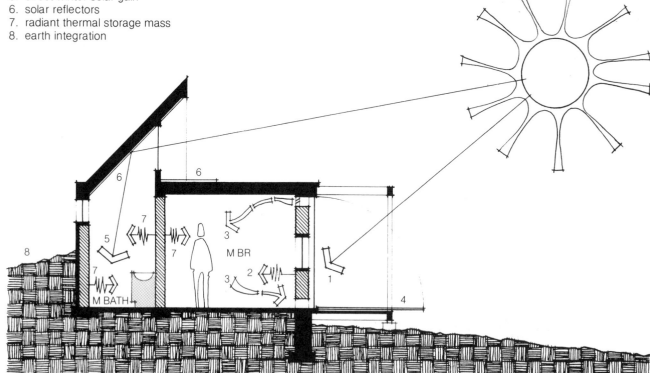

Section through master bedroom

10 FT
3 M

The solar aspects of the house are a constant reminder of the weather. The Trombe walls collect heat during the day and begin to radiate it into the house at about sundown, when it is most needed. The water barrels in the solarium collect and radiate heat in much the same way. They keep the solarium at a much more moderate temperature than we experienced before the barrels were installed.

We live very informally with most of our activity centering around the kitchen. This has become the heart of the house, since we entertain and both of us spend a lot of time cooking. When we entertain, guests are also naturally drawn into the kitchen, so whoever is cooking is not isolated from the others. The openness of the house greatly contributes to a free-moving flow of conversation from room to room.

Since moving into the house, we've noticed very little fluctuation of indoor temperature. This is partly due to the addition of special Venetian blinds on the upper solarium windows: one side is black and the other silver-colored to absorb or reflect sunlight, allowing more control. A project for the future is to add insulating reflective shutters outside the glass of the Trombe walls. These will keep heat out in the summer when they are closed, and reflect more solar gain in cold weather, when they are open during the day.

For us, the house is many things: a passive solar house, a comfortable home to return to, an economical house to operate, and a unique expression of our values.

1. roof support for summer shading
2. insulated vent (open)
3. sun-control blind
4. sloped glazing
5. insulated vent (closed)
6. water drums
7. insulated shutter and reflector

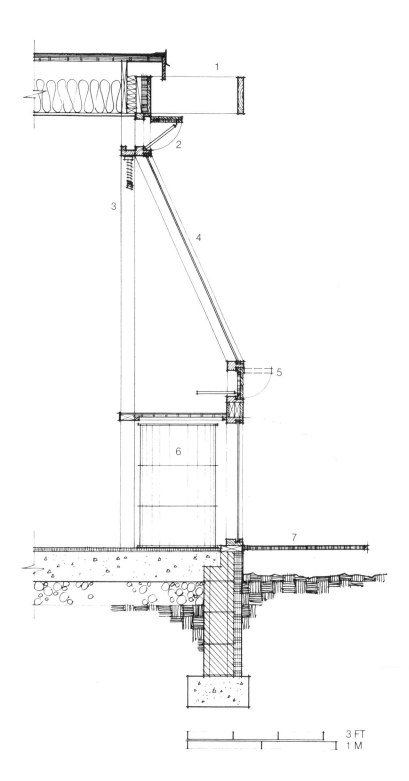

3 FT
1 M

Detail of sunroom

Project Data Summary

Project Information

Project: Dehkan residence
          Andover, New Jersey
Architect/designer: Kelbaugh & Lee—Doug Kelbaugh
                     Princeton, New Jersey
Builder: Keith Andrews
          Newton, New Jersey

Climate Data

Latitude	40.0°N
Elevation	250 FT (76m)
Heating degree days	5,696
Cooling degree days	927
Annual percent possible sunshine	58%
January percent possible sunshine	50%
January mean minimum outdoor air temperature	24°F (−4.5°C)
January mean maximum outdoor air temperature	41°F (5°C)
July mean minimum outdoor air temperature	64°F (17.5°C)
July mean maximum outdoor air temperature	86°F (30°C)

Climate features: harsh winters; warm, humid summers

Building and System Data

Heated floor area	1,704 FT² (158.5m²)
Solar glazing area	
Direct gain	185 FT² (17m²)
Solarium	194 FT² (18m²)
Trombe wall	177 FT² (16.5m²)
Thermal storage heat capacity	
Trombe wall	5,770 Btu/°F
Tiled concrete floor	13,860 Btu/°F
Masonry walls	16,380 Btu/°F
Water storage	4,790 Btu/°F

Performance Data

Building load factor	5.5 Btu/DAY°F FT²
Auxiliary energy (heating)	3.0 MMBtu/YR
Auxiliary energy (cooling)	1.2 MMBtu/YR
Solar heating fraction	92%
Night ventilation cooling fraction	94%

Rooms with "legs" extend beyond the house to create outdoor living spaces.
Peckham residence, Columbia, Missouri.

A three-story Trombe wall captures and releases solar heat. Peckham residence.

Architect Nicholas Peckham has a commitment to his work gained in the best way—by living in his own passive solar home. This house is a personal summary and statement of his work. The site is in a traditional middle-class neighborhood in Columbia, Missouri. The existing homes are what anyone might have aspired to years ago—or even today. Although this house is unlike the other homes in the neighborhood because it is solar, it complements them visually.

Structurally, it is set up on a four-foot building module to minimize construction waste. From the outside, the house appears larger than it really is. Since the site slopes slightly from west to east, half levels were created. As one enters, there is a choice between descending half a flight to the exercise area or ascending half a flight to the living area. In total, there are five levels within this three-story structure. Additionally, rooms and decks have "legs" that extend beyond the main shell of the structure to provide a protective entry on the south and a bedroom deck and television room on the north.

The solar system takes three forms: a Trombe wall, a greenhouse, and some direct gain. The Trombe wall extends from the exercise room on the lower level to the living/dining area and upward to the master bedroom—three stories in all. The greenhouse begins at the foyer and extends to the art gallery above. The bedroom to the west is linked thermally to the greenhouse by a duct system. Secondary direct gain makes up the remainder of the heat in this area.

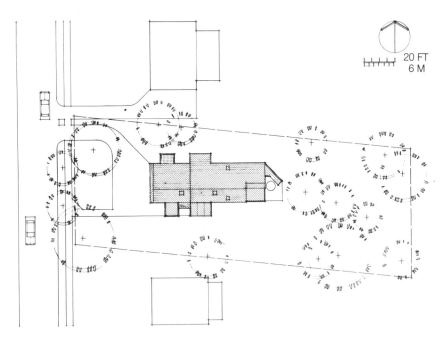

20 FT
6 M

Site plan

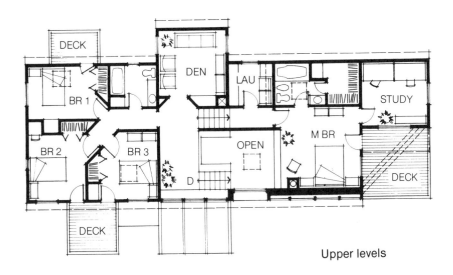

Upper levels

Mid levels

Lower level

10 FT
3 M

Heating of all spaces is accomplished quite well with a combination of systems. Because the five levels open onto one another, free air movement is allowed. A winter feature provides mixing of warm stratified air with cooler air by the auxiliary-system fan to minimize the need for auxiliary heating. In summer, cooling is accomplished by a ceiling fan (to remove stratified air) and natural ventilation. During extreme conditions, an auxiliary electric air conditioner cools and dehumidifies. Seasonal expansion of the living area is possible with the screened porch to the east.

1. indirect winter solar gain
2. radiant heat from greenhouse thermal storage
3. warm air from greenhouse
4. return air to greenhouse
5. duct to bedroom zone
6. direct winter solar gain
7. solar water heater collectors

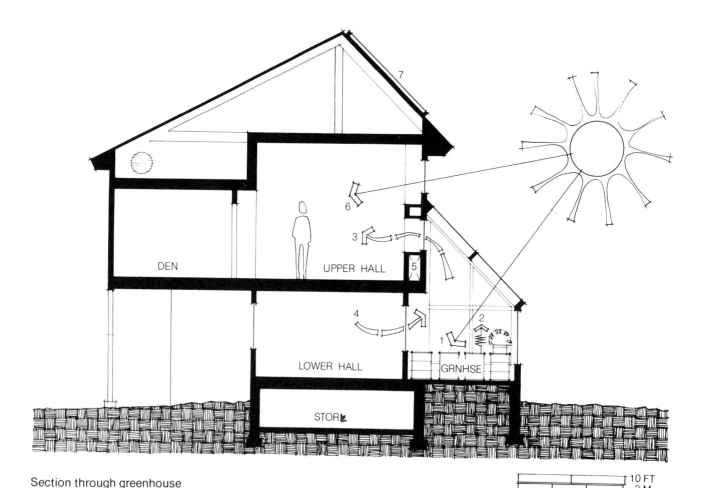

Section through greenhouse

10 FT
3 M

Combining passive solar systems with balanced space planning inside a tight weatherskin, Peckham has created an energy-conscious environment for his family. The beauty of this design is that it is a contemporary statement that satisfies the owners and adds vitality to the neighborhood.

1. indirect winter solar gain
2. radiant heat from Trombe wall
3. natural convection
4. isolation dampers
5. summer vent
6. energy-efficient wood stove
7. north buffer zone
8. skylight

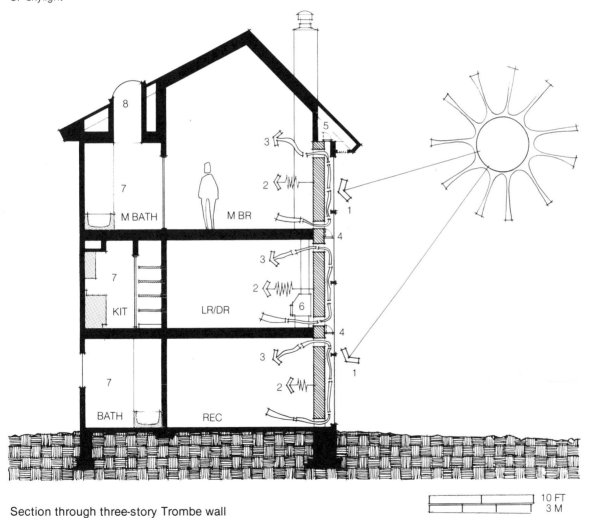

Section through three-story Trombe wall

10 FT
3 M

The openness of the interior makes for dramatic spatial experiences. Peckham residence.

From the Architect/Inhabitant

The first six months of living in this home have produced a number of interesting experiences. The house cost a great deal more than any other this family has lived in, but the heating expenses for the whole winter were less than a typical month in the small college faculty house we moved from. Many people stop by to look, and quite a few knock on the door and ask to come in. The house is on a lot whose long axis is east-west, with a beautiful view to the east backyard. This results in a house that faces the street, unlike the other homes in this old neighborhood. The openness, both vertical and horizontal, of the interior spaces makes for dramatic spatial experiences, but creates some acoustical separation problems.

The size, shape, color, and feel of this house are different and better here than in anything we have lived in before. Even if it weren't an energy-conscious design, the visual delight would be remarkable. Living here is a real joy because the passive solar conditioning is always comfortable and because the architecture is inspiring.

Our children are in fifth, ninth, and tenth grades; they comment, "I like it a lot." More important are the ways in which their social behavior has been affected: They spend more time at home, they have many more visitors and guests, they discuss energy issues with their friends, they write energy reports for their school assignments.

Project Data Summary

Project Information

Project: Peckham residence
Columbia, Missouri
Architect/owner/builder: Nicholas Peckham, AIA
Columbia, Missouri

Climate Data

Latitude	38.8 °N
Elevation	746 FT (227.5m)
Heating degree days	5,083
Cooling degree days	1,269
Annual percent possible sunshine	63%
January percent possible sunshine	52%
January mean minimum outdoor air temperature	21°F ($-$6°C)
January mean maximum outdoor air temperature	38°F (3.5°C)
July mean minimum outdoor air temperature	67°F (19.5°C)
July mean maximum outdoor air temperature	87°F (30.5°C)

Climate features: cold winters; dry summers

Building and System Data

Heated floor area	3,500 FT^2 (325m^2)
Solar glazing area	
Trombe wall	416 FT^2 (38.5m^2)
Sunspace	240 FT^2 (22m^2)
Direct gain	268 FT^2 (25m^2)
Thermal storage heat capacity	
Masonry Trombe wall	12,480 Btu/°F
Concrete sunspace floor	1,440 Btu/°F
Water drums	25,670 Btu/°F

Performance Data

Building load factor	5.4 Btu/DAY°F FT^2
Auxiliary energy (heating)	47.4 MMBtu/YR
Auxiliary energy (cooling)	12.2 MMBtu/YR
Solar heating fraction	49%
Night ventilation cooling fraction	58%

A two-story mass wall dominates the narrow face of the south apartment. Baker residence retrofit, Berkeley, California.

A solar retrofit, this duplex shows how a dilapidated urban shoebox can be converted into a treasure with solar style. Architect David Baker worked hard and had fun designing playful habitats out of this Berkeley derelict. The knowledge gained through designing and living in his own passive solar home lends the most insight to his future designs.

This building presented a myriad of unusual design restraints that Baker managed to change from liabilities into assets. Careful adaptation of passive and active solar elements minimized the energy consumption of space heating, domestic water heating, lighting, and the indispensible California hot tub. The solar elements were combined and squeezed in where possible: a two-story mass wall dominates the narrow face of the south apartment, a thin band of direct-gain glazing surrounds the sloping active collectors over the north apartment bedroom, a translucent roof over the north-unit bath floods the space with daylight, and high direct-gain glazing heats the northern portion of the building. The architectural treatment of each solar feature adds to the flair of the total design. In this case, the investment paid off with a handsome architectural and economical return.

From the Architect/Inhabitant

Built in 1911 as a grocery store with attached apartment, the existing structure completely filled the 14-by-78-foot (4-by-23.5m) corner lot. I had been walking by this funny little building for years and had thought it would make an interesting renovation project. I finally gave in and bought it. The grocery store had been converted to an apartment in the 1950s, so the building was now a duplex. Since a great deal of major structural repair was needed and the available space was absolutely limited by the property lines, I decided to design with functional and esthetic considerations tak-

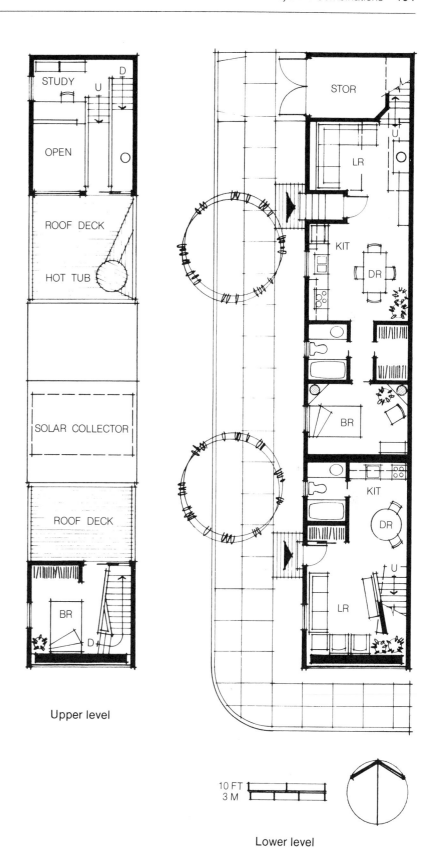

Upper level

Lower level

10 FT
3 M

Retrofitting altered a dilapidated shoebox into a solar treasure. Baker residence retrofit.

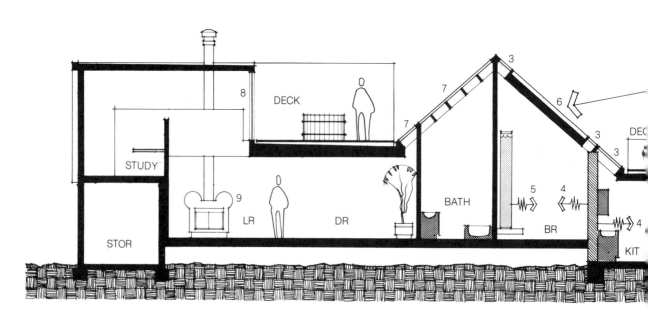

Section

ing precedence over the existing structure.

The final design was a response to a number of problems and opportunities, of which solar energy was only one consideration. Single elements of the design frequently evolved a solution satisfying more than one concern. For example, a raised gable between the two apartments provides a mounting surface for the solar collectors and serves as a visual and acoustical barrier between the private roof decks. These decks, one with a solar-heated hot tub, make up for the absence of yard space. South-facing sliding glass doors to one of these decks provide a major passive solar-collection area for the northern apartment. Since no windows were possible on the east wall (because of fire codes), strip overhead glazing brings daylight into the building and allows some direct gain.

Double-height rooms, including the 8-by-8-foot (2.5-by-2.5m) living room, visually expand the limited floor area while allowing hot summer air to stratify and be exhausted through vents. Insulation to meet the present energy conservation standard was added, and new double-glazed window units were installed. The wooden windows have superior infiltration and insulating characteristics. In addition, they allow the use of blinds between the double glass, which enhances their insulating value and allows complete privacy on the street side.

An active solar domestic hot-water system serves both units and provides heating for the hot tub. I decided to integrate the collector with the roof, using corrugated fiber glass to cover both the collectors and direct-gain openings.

The mass wall provides more than 90 percent of the space heating for the

south apartment. The mass consists of 8-inch-thick (20.5cm) concrete block, plastered solid with concrete and finished with half-inch-thick (1cm), red-clay quarry tile on the interior surface. The glazing is corrugated fiber glass. It is attached to wood furring strips mounted directly on the concrete block. The mass wall is held in eight inches from the top and the side walls, creating a strip window that provides direct gain at the edges and eliminates the need for air circulation between the glazing and the wall. The strip window also provides an architectural separation between the mass wall, the side walls, and the ceiling, visually and structurally floating the heavy masonry element. The effect is quite beautiful, both during the day, when the strip of sunlight acts as an internal sundial, and at night, when the interior lights illuminate the strip to the outside.

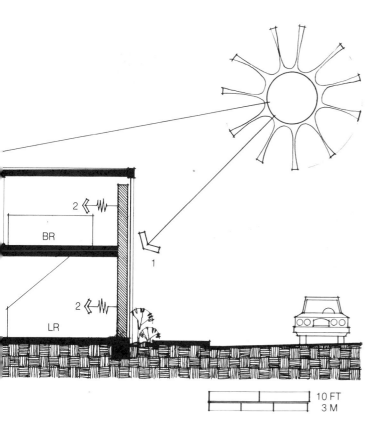

1. indirect winter solar gain
2. radiant heat from mass wall
3. skylights
4. radiant heat from thermal mass
5. radiant heat from water tubes
6. solar water heater collectors
7. translucent roof
8. clerestory to living area
9. energy-efficient wood stove

10 FT
3 M

The upper roof deck compensates for tight site. Baker residence retrofit.

If I were doing this wall again, I would attempt either to increase the number of glazing layers, or to provide movable insulation between the glazing and the wall. Though the wall works perfectly when the sun shines, during long cloudy periods the mass becomes a large, cold surface. In spite of this problem, the backup heat for the entire unit is supplied by one small electric-resistance heater. Energy costs have been minimal in the south studio apartment, with an average electrical bill for space heating, water heating, cooking, and lighting of under $10 a month.

The north apartment has no significant structural thermal mass, as there was no opportunity to integrate it. Thermal mass was added with two 10-foot-high (3m), 18-inch-diameter (45.5cm) fiber-glass tubes filled with blue-colored water. Bracing water tubes so they won't fall during an earthquake is a necessity in California; I didn't want to become the subject of a syndicated newspaper story entitled, "Solar Home Owner Killed in Bed by His Thermal Mass." I had heavy steel triangular brackets fabricated. The popular conception of a water thermal storage system as a bunch of ugly black drums is not fulfilled here; the translucent tubes are almost luminescent, and visitors have mistaken the installation for a sculpture.

There is no doubt that the passive solar features added to the construction cost. However, they also add an esthetic richness that would be lost without them. The total renovation was done within a budget of $40 per square foot, and the immediate resale value of the entire renovated structure was 50 percent higher than the total investment.

Translucent water-storage tube is a sculptural element as well. Baker residence retrofit.

Project Data Summary

Project Information

Project: Baker residence retrofit
   Berkeley, California
Owner/architect/builder: Sol-Arc—David Baker
      Berkeley, California

Climate Data

Latitude	37.9°N
Elevation	345 FT (105m)
Heating degree days	2,909
Cooling degree days	128
Annual percent possible sunshine	70%
January percent possible sunshine	52%
January mean minimum outdoor air temperature	40°F (4.5°C)
January mean maximum outdoor air temperature	56°F (13.5°C)
July mean minimum outdoor air temperature	54°F (12°C)
July mean maximum outdoor air temperature	76°F (24.5°C)

Climate features: mild, rainy winters; cool summer evenings

Building and System Data

Heated floor area	1,200 FT² (111.5m²)
Solar glazing area	
Mass wall	180 FT² (16.5m²)
Thermal storage heat capacity	
Tiled concrete floor	12,000 Btu/°F
Concrete masonry mass wall	5,400 Btu/°F
Water storage tubes	2,205 Btu/°F

This earth-integrated structure is burrowed in the desert. Hoffmann residence, Tempe, Arizona.

The greenhouse off the living area provides light, heat, and humidity. Hoffmann residence.

The southwest desert is the stage for strong design statements, both natural and man made. The landscape and climate combine to create an environment which demands specialized adaptation. Plants and animals have evolved forms, textures, and living patterns which help them survive the harsh forces of nature. The desert climate is typified by mild winters with bright, sunny days. The hot season is long, with intense, dry, sun-bleached days and clear, starry, cool nights. Daily temperature swings of 30–40°F (−1–4.5°C) are not uncommon in any season. Strong spring winds, blowing sand, intense solar radiation, and dryness are additional features.

Architectural designer James Hoffmann has taken some tips from nature to beautifully adapt strong, logical, man-made forms to this rugged climate. This earth-integrated structure uses a double-vaulted roof, sod roofs, ventilating light monitors, and seasonal sun shades. Most of the passive solar features of the design address cooling, which in this area is more than twice the heating need.

The double-vaulted roof is oriented with its long axis north-south. The long metal outer vault is set off from the subroof to both shade and allow free air passage between. The dark-colored outer metal skin heats up during the day, creating an upward convective air current between roofs; this helps to cool the inside roof surface, significantly reducing the need for cooling. Dampers at the top of the vault can be opened to allow warm stratified interior air to escape. This stack effect is augmented by the motion of the outside heating air moving between the roofs, helping to induce interior air up and out. Air can also be drawn from outside shaded gardens, the evaporative cooler, or through a rock bed.

In winter, the west half of the dark vaulted roof acts as an unglazed solar collector. A fan draws warmed air from between the double roof down through the subfloor rock bed. This acts as long-term thermal storage, with both radiant and forced-air distribution to the interior when needed. The vault collector meets most of the heating needs not met by the direct-gain and sunspace systems.

During hot spells, an evaporative cooling system is used to cool the space directly, an effective cooling method in this relatively dry climate. At night, the evaporative system can significantly cool the subfloor rock bed. During the day, interior air is circulated through this coolth sink, effectively reducing its temperature.

The earth roofs and earth-bermed walls act as a thermal buffer for much of the structure. The roof and walls are heavily insulated to a few feet below finish grade. The uninsulated concrete walls and floor slab which contact the earth take advantage of its subsurface coolness. This mass acts as a thermal flywheel, absorbing or releasing heat at a rate much slower than outside temperature changes. This time lag dampens the effect of the high outside daily temperature variations, resulting in only moderate interior temperature changes.

This home features a number of desert design strategies: the roof monitors face northwest, allowing diffuse daylight and ventilation; an active solar domestic water heater is integrated into the southern side of the monitors; all vertical glazing is carefully recessed or shaded by awnings to avoid direct summer gain; most glazing is operable, to allow night ventilation and cooling. The rust-colored steel roof, cement stucco, and planted earth were selected to resist the deteriorating effects of wind, sand, and sun, and require practically no maintenance in this desert environment.

From the Designer/Inhabitant

This residence is the first of a six-lot planned unit development. The planning includes common green space and a garden/orchard area. An off-street alley provides vehicle access and parking. Simple measures such as reduced paving on the interior of the lots, lush landscaping near the homes, and native desert landscaping around the subdivision perimeter are unusual for developments in sprawling greater Phoenix.

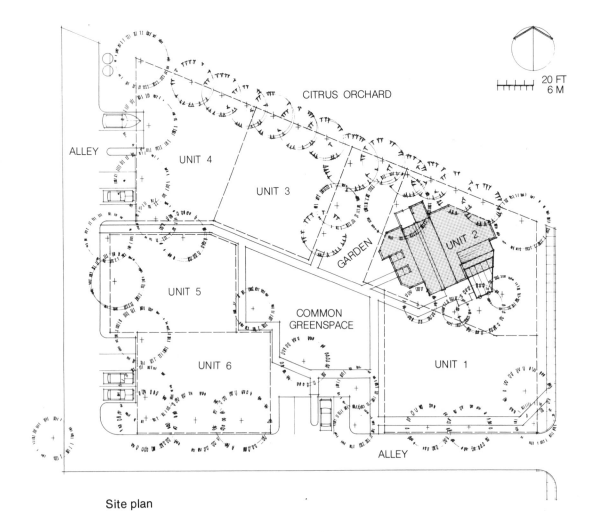

Site plan

Deeply recessed windows are shaded from sun, while vault vent allows warm stratified air to escape. Hoffmann residence.

We have established the following goals for the first residence:

1. Demonstrate suitable desert architectural design and generate interest from commercial developers.
2. Reduce energy consumption, with special emphasis placed on cooling.
3. Introduce consumers to various energy-conscious design features and concepts, efficient land use, water-conservation devices, and indigenous landscaping.
4. Collect performance data for future design and development (eighty areas of the house are being monitored).
5. Provide a teaching aid and demonstration tool for area schools and community organizers.
6. Test public marketability and response by loan institutions.

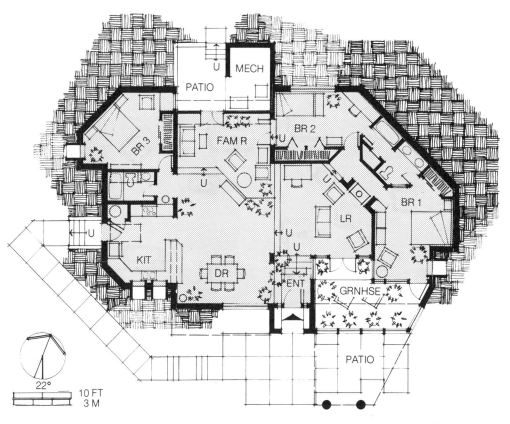

Floor plan

Massive interior walls and floors store heat and coolness. Hoffmann residence.

Design features include the following:

1. Optimum orientation—solar-gain surfaces twenty-two degrees east of south for morning heat pickup and shading in the afternoon
2. Landscaping—low-water-consuming native plants, high trees to northwest for shading in the afternoon
3. Earth integration—approximately 68 percent of total building envelope has contact with earth (rather than with atmosphere)
4. Selective shading—fixed overhangs and louvers, awnings, tube windows for difficult east and west orientations
5. Main vault roof—an unglazed collector in winter and double-shade roof in summer
6. Thermal mass—earth tempering, masonry walls, tile floors on concrete slab
7. Rock bed—twenty tons of rock, providing long-term thermal mass for winter heat and summer cooling.
8. Domestic hot water—active solar collector architecturally integrated with roof monitors
9. Backup heat—rock bed and wood-burning stove provide all supplementary heat.

With the completion and occupancy of this prototype, the exciting process of performance observation and reaction of commercial developers and general public begins. I am optimistic about the acceptability of energy-efficient housing. People seem to be ready for it.

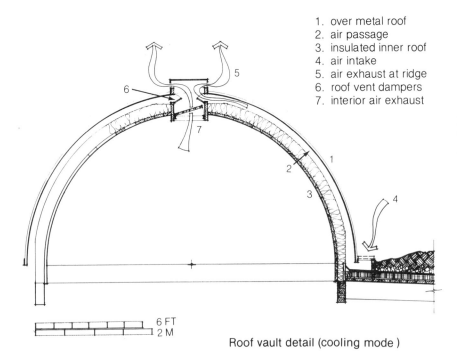

1. over metal roof
2. air passage
3. insulated inner roof
4. air intake
5. air exhaust at ridge
6. roof vent dampers
7. interior air exhaust

6 FT
2 M

Roof vault detail (cooling mode)

1. earth integration
2. radiant thermal storage mass
3. solar water heater collectors
4. solar storage tank
5. vent and light monitors
6. recessed glazing

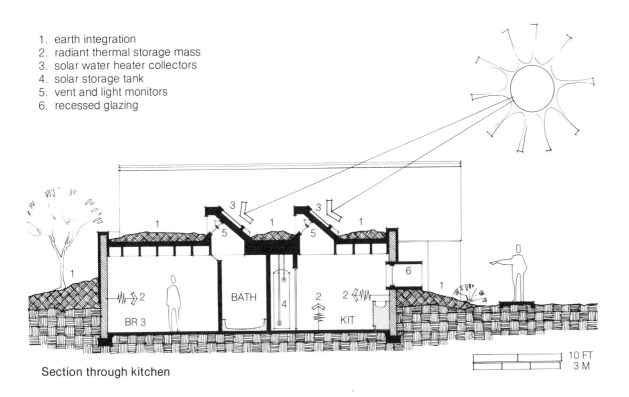

Section through kitchen

1. indirect winter solar gain
2. natural convection
3. vent and light monitor
4. shading
5. earth integration
6. radiant thermal storage mass
7. energy-efficient wood stove

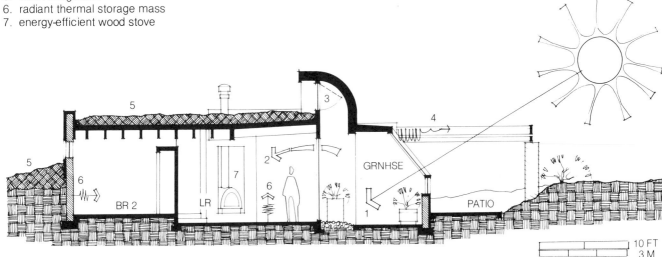

Section through greenhouse

Project Data Summary

<u>Project Information</u>

Project: Hoffmann residence
Tempe, Arizona
Designer/builder: ECOgroup—James Hoffmann

<u>Climate Data</u>

Latitude	33.5 °N
Elevation	1,117 FT (340.5m)
Heating degree days	1,532
Cooling degree days	4,000
Annual percent possible sunshine	86%
January percent possible sunshine	77%
January mean minimum outdoor air temperature	38° F (3.5°C)
January mean maximum outdoor air temperature	65° F (18.5°C)
July mean minimum outdoor air temperature	78° F (25.5°C)
July mean maximum outdoor air temperature	105° F (40.5°C)
Climate features: hot, arid summers; mild winters	

<u>Building and System Data</u>

Heated floor area	1,750 FT² (162.5m²)
Solar glazing area	
Direct gain	250 FT² (23m²)
Greenhouse	172 FT² (16m²)
Thermal storage heat capacity	
Masonry retaining wall	23,760 Btu/° F
Rock bed	10,240 Btu/° F
Interior masonry wall	1,560 Btu/° F
Tiled concrete floor	11,140 Btu/° F
Greenhouse tiled concrete floor	2,750 Btu/° F
Greenhouse masonry wall	3,860 Btu/° F

<u>Performance Data</u>

Building load factor	7.1 Btu/DAY° F FT²
Auxiliary energy (heating)	0.0 MMBtu/YR
Auxiliary energy (cooling)	17.3 MMBtu/YR
Solar heating fraction	100%
Night ventilation cooling fraction	82%

Chapter 6

HYBRID SYSTEMS

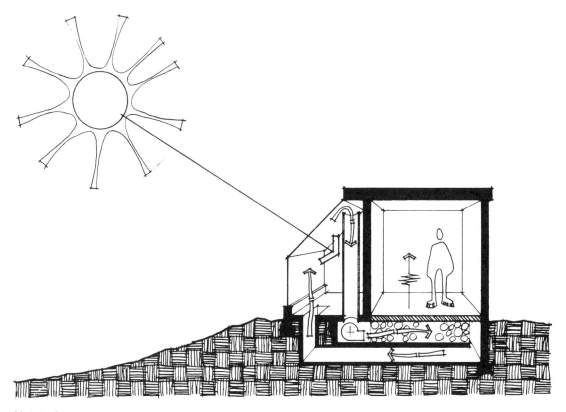

Hybrid Systems

Although pure passive solar systems or combinations can be designed to work well in most climates, designers sometimes choose to incorporate mechanical features to assist in the primary collection, storage, or distribution of thermal energy. When the operation of the passive system relies on mechanical assistance, it is classified as a hybrid design. Through the use of a fan or a pump, a hybrid system offers the following advantages:

1. Improved collection of solar energy by effectively moving heat to storage mass
2. Removal of excess solar heat to prevent overheating
3. Effective long-term thermal storage by stockpiling solar heat when available
4. Elimination of hot or cool areas by balancing distribution
5. System automation, reduced occupant participation
6. Increased opportunity for integrating passive solar and auxiliary systems

An example of a hybrid design is a passive solar greenhouse which collects a large amount of solar energy, yet contains no thermal mass. A thermostatically controlled fan draws off the heated air to rock-bed storage located well within the building. The stored heat may slowly conduct through a floor slab, then radiate to warm a north room. Thermal storage is mechanically charged while distribution is passive.

Nestled in an apple orchard, this solar showpiece is at one with its environment. Thompson residence, Westford, Massachusetts.

Central sunspace brings outdoors in. Thompson residence.

It is a delight to see complete and competent architectural design that incorporates emerging technological concepts. Massdesign, through the team effort of architects Tudor Ingersoll, Gordon Tully, and Mollie Moran, has explored hybrid solar designs and come up with a fine architectural statement. Concern with and insight into all architectural aspects from concept to detail produced this functional and beautiful place to live. A surprisingly moderate construction cost was due to many factors, including the use of common materials and a simple architectural form.

This house's solar system is similar to other successful designs which thermostatically circulate air warmed by excess solar gain or auxiliary equipment to a thermal storage mass. Circulating the warm air to both the rock bin and heat exchanger from the oversized active water tank is unique and seems to function well. Slight operational modifications and the use of new products and controls offer to make this type of system an acceptable approach to most any home buyer.

The enthusiasm to create a hybrid system was second only to the client's needs and desires. The operational aspects of the system were thought out prior to construction and then followed up and reevaluated after occupancy. The final result is a fine piece of residential architecture as well as a positive learning experience for all involved. In the end, it appears that the owners are more accepting of this new type of climatized environment than the architects had anticipated—a pleasant surprise!

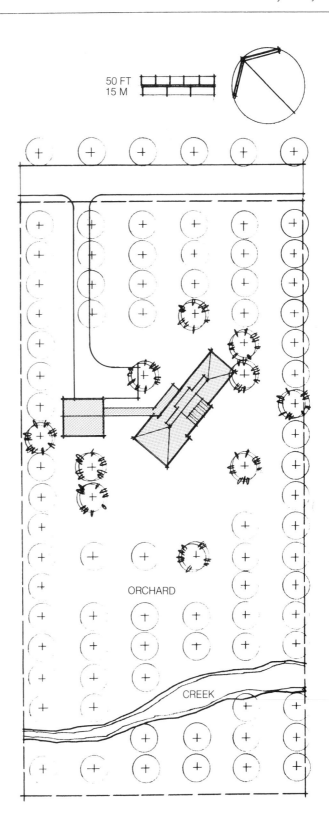

50 FT
15 M

ORCHARD

CREEK

Site plan

From the Architect

The Thompson house was very definitely a team effort—the architects provided design expertise and analysis and the Thompsons provided patience and enthusiasm. They had no idea when they started the project that they would end up with a showpiece, but they were extremely open-minded and willing to learn, and accepted enthusiastically many advanced architectural concepts which came along with the solar ones.

The Thompsons wished to provide minimal quarters for visiting children. They had a site with good solar exposure in an apple orchard and a very simple request: they wanted a house with minimal auxiliary heating. Our basic architectural response was a combination of the following:

1. A house stretched out east-west with all major rooms facing south.
2. The use of direct-gain passive solar heating as a primary energy source.
3. Oversized solar water heater for domestic hot water and auxiliary space heating.
4. Thermal storage within the building, but with no attempt to provide all the necessary thermal storage within the living space.

Given these requirements, a natural scheme developed. All the living spaces were put on the first floor, with two small sleeping rooms tucked under the ridge of the roof and left cool during the winter to cut down on heat loss. The large sheltering hip roof was a response to the small second floor and the large first floor, and it nestled very nicely into the orchard. In a very long plan such as this, it is necessary to provide some kind of passageway between the two ends of the house; it was placed along the south wall. The entry and kitchen look out over it, and part of it is expanded to create the dining-room sunspace. The entrance, bathroom, stairway, and other support facilities are all located in the middle of

The passageway along the south wall connects the east and west ends of the house. Thompson residence.

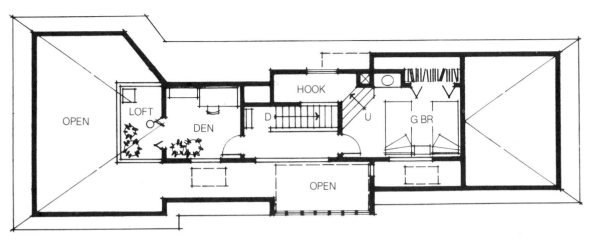

Loft level

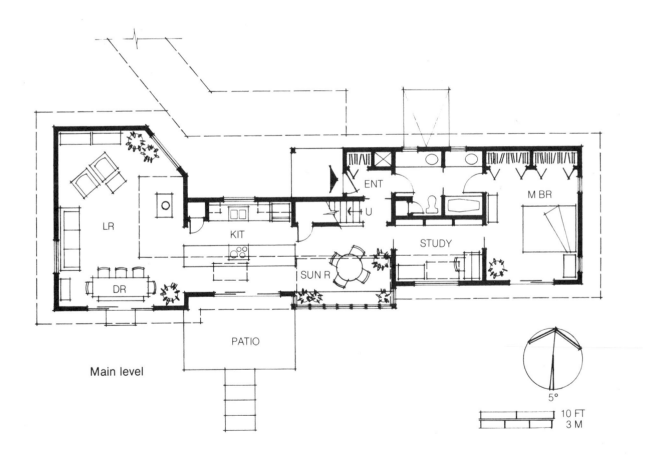

Main level

5°

10 FT
3 M

the plan, at the north. Upstairs, the bedrooms can open to the lower space by means of shuttered interior windows, from which two skylights can be opened for ventilation. Architecturally, the concept resembles a large hip-roofed barn, within which is suspended a long second-story element. The roof is visible from inside at both ends of the plan.

The basic hybrid system initially seemed simple, but proved to be extraordinarily complicated—the attempt to describe ahead of time what

the control sequence might be was mind-boggling. A large rock bin and the coil from the active solar water-heating system are coupled in series with the house's conventional forced-air system. When the furnace fan is turned on, air is drawn through a high return at the ridge and a low return at the entry. It then passes over the heat-exchanger coil (which is either on or off, depending upon the temperature of the active water storage tank) and through the rock bin, ultimately supplying air registers at the windows.

The coil supplies heat to the air stream if the stored water is hot from the solar collectors. The rock bin stays at basically the same temperature as the house; if the house gets cold, so does the rock bin. Then the family relies on a wood stove to keep them warm. We explained that the air-conditioning mode must be operated whenever there was excessive heat in the living space, at which time the cool discharge air would not be uncomfortable. It appears that the Thompsons have adjusted to this system and in

Dining sunspace juts out from the central passageway. Thompson residence.

fact enjoy it, especially after we went back and quieted the noisy fan and reduced the discharge velocities in some grills.

If we had it all to do over again, we would use a reversible-flow rock bin that charges and discharges from the same end, so that the warmest storage in the bin is always used. We would also incorporate more mass within the living space, although the reason that the Thompsons are so comfortable may have to do with the fact that we hit it just right by putting about half the thermal mass in the living space and half in a remote space. Finally, we would insulate more, triple-glaze the windows, and improve on the lightweight insulating shutters that were used. What we would not change are the architectural features, the incorporation of a glazed sun-space within the living space, and the use of the hybrid system.

1. winter solar gain
2. radiant thermal storage mass
3. high air to rockbed
4. low air to rockbed
5. fan
6. rockbed
7. solar water heater collectors
8. solar storage tank
9. heat exchanger coil
10. sun-control screen

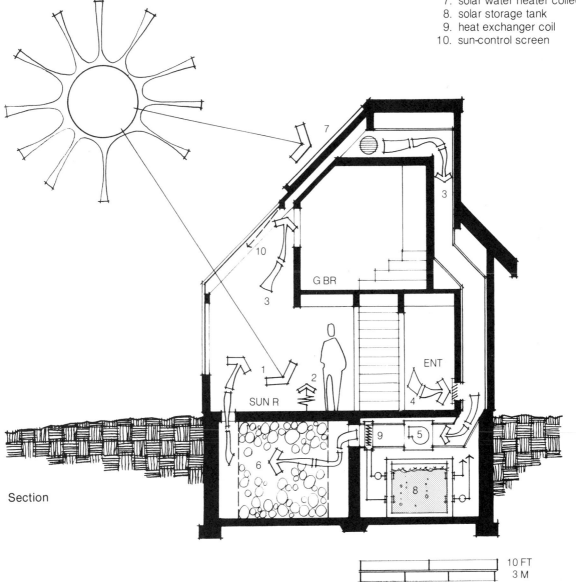

Section

10 FT
3 M

From the Inhabitant

Because this house is designed partly to absorb solar energy directly, it is necessarily open to our natural surroundings. We feel that we live in an old orchard, surrounded and sheltered by living spaces, rather than in a house surrounded by an old orchard. These living spaces can be open or made snug by means of insulated shutters. We have a wonderful feeling of control over living conditions in the house and of independence from outside factors such as power outages or fossil-fuel shortages.

It will always be exciting to look up from the breakfast table in the sunspace and see a flock of geese directly overhead, or to pass through the same space at night and see the stars so easily. Being surrounded by sunlight and living plants can really make a February day in New England more enjoyable.

There has been an undeniable change in our lives. Indoor activities are determined by what area has been made most comfortable by the sun. We feel that our lives are much more harmonious with the rhythms of the seasons. New hobbies such as indoor gardening and star gazing have been added, but have not replaced other of our activities.

In addition to the esthetic values, we find that this is a very healthful environment. The humidity stays at about 50 percent during the winter. We generally feel marvelous—only one unshared cold has been suffered in almost two years. In summer, the southern exposure works to our advantage by opening to the prevailing southwest breezes. The deep overhang and the greenhouse sunshades create the effect of an airy pavilion.

Project Data Summary

Project Information

Project: Thompson residence
    Westford, Massachusetts
Architects: Massdesign—Tudor Ingersoll, Gordon Tully, Mollie Moran
    Cambridge, Massachusetts
Builder: Donald Bourgeois
    Waltham, Massachusetts

Climate Data

Latitude	42.8°N
Elevation	160 FT (48.5m)
Heating degree days	5,900
Cooling degree days	565
Annual percent possible sunshine	55%
January percent possible sunshine	48%
January mean minimum outdoor air temperature	18°F (−7.5°C)
January mean maximum outdoor air temperature	36°F (2°C)
July mean minimum outdoor air temperature	62°F (16.5°C)
July mean maximum outdoor air temperature	84°F (29°C)
Climate features: variable, with seacoast influence	

Building and System Data

Heated floor area	1,300 FT² (120.5m²)
Solar glazing area	
Direct gain	350 FT² (32.5m²)
Active solar collector	111 FT² (10.5m²)
Thermal storage heat capacity	
Tiled concrete floor	8,000 Btu/°F
Rock bin	6,500 Btu/°F
Active-system water-storage tank	4,000 Btu/°F

HAUMAN RESIDENCE: Guilford, Connecticut

Attention to craftsmanship resulted in this masterful solar structure. Hauman residence, Guilford, Connecticut.

The virtually windowless north side acts as a buffer between the house and inclement weather. Hauman residence.

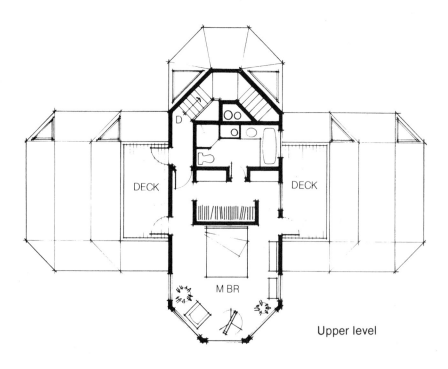

Upper level

Few designers in today's fast-paced world are able to guarantee the incorporation of fine materials and excellent craftsmanship in their projects but Conger and Lytle in Guilford, Connecticut, have managed to revive the art of the master builder in a twentieth-century way. This collaboration by two young designer/builders brings together modern architectural training and journeyman construction to create high-quality, energy-efficient buildings. Their designs interweave the best qualities of both traditional building and design techniques with modern materials and systems. Their residences are a fusing of site-generated forms, on-site design, and, often, owner input during construction. Working drawings are minimized; these designers use conceptual sketches and scale models to work out the design as it evolves. At each stage of conceptualization and construction, energy-conserving methods and materials are integrated to assure efficient space conditioning.

The master-builder tradition demands that all aspects of the building be masterful. Too often in "modern" architecture, comfort is sacrificed for style, effect, or expression, and the basic elements of physical comfort are overlooked. This home is a fine example of comfort, energy conservation, and sense of place.

From the Designers

The first and foremost response we seek when designing a dwelling is the subtle human emotion which associates the form of a building with one's conception of "home" or dwelling place. Each design integrates a consciously developed energy system, but never allows the emotional esthetic of home to be secondary. That the home incorporates solar techniques is assumed when we begin a design.

We generate each design concept from the site as well as the owner's needs and images of home. The site may suggest a tall building that rises out of the trees to capture light and breeze, or to secure a particular view. The building may want to stretch horizontally to inhabit a field or meadow. One face of the home seeks exposure to light, and therefore energy. This face usually has expanses of windows, areas of glass which must be delicately placed to achieve a balance of light. In fact, one of the most pleasing elements of this type of design is the vocabulary of light and shadow.

The subject of detailing introduces another critical criterion in our vocabulary of design: fine craftsmanship. The selection of materials and the manner in which they are utilized and applied completes the relationship of concept and realization. Without hand-crafting by either owner or master builder, the home remains unnamed, unmarked, and anonymous. The doors, the gutters, all details must receive special consideration. For most of our clients, a home is the single largest investment of their lives. It is important, therefore, that their habitat be well constructed and demonstrate sensitivity in the elements which they will experience daily.

The energy system is critical to this living experience. With effective performance, the home responds to its occupants' daily comfort requirements. The direct-gain passive solar system is not necessarily the sole solution to effective comfort control. A hybrid system of air distribution is usually important.

On the first floor in this house, sunlight directly strikes a tiled concrete slab which caps a 21-cubic-yard (16m³) low-temperature rock storage bed. As in most passive systems, the house itself is used as a collector. Excess heat captured by the house rises to the upper level, where it is cycled into the rock storage, later to permeate upward through the slab into

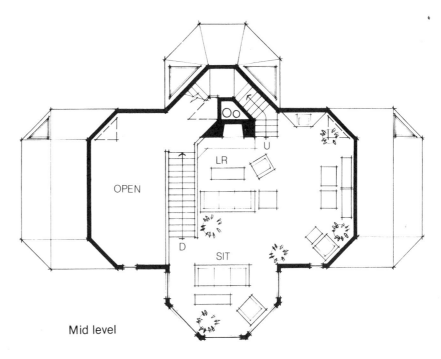

Mid level

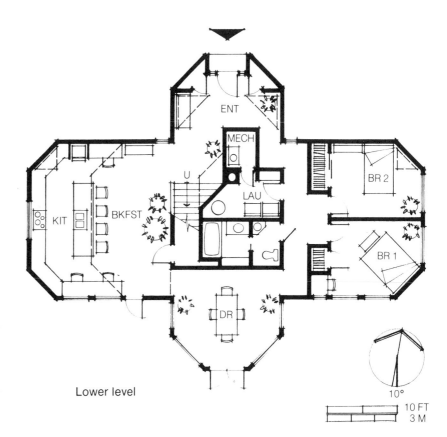

Lower level

the living areas. This recycling is accomplished with a thermostatically controlled fan that blows air into the rock storage when it reaches a high enough temperature. This is an actively charged, passively discharged hybrid system. A windowless, full-height stair tower on the north side allows heat flow from the lower areas of the house to the high air-collection point. The virtually windowless north side buffers the house from inclement weather, and high skylights perforate the north roof to allow light into the house's rear spaces.

The house has an oil furnace and fireplace for auxiliary winter heating. In summer, all glazing areas open to permit breezes, and night air blown through the rock storage provides additional daytime cooling.

From the Inhabitant

We have lived in this home for about a year; we love it, not because it is a solar house, but because of the site and design. We fell in love with it after seeing a newspaper ad for an open solar house. Since we were planning to build a solar home of our own, we decided to see what was being built. What started as curiosity turned into obsession, and we went back time and time again to look at and experience this house, and we eventually purchased it.

The interior spaces are perfect for us. The high ceilings, open spaces, and fine craftsmanship suit our tastes very well. Living here has modified our lives somewhat. We entertain more often, and when we have friends over,

the rug is rolled back in the upper-level living room for dancing and the kitchen becomes the bar and buffet—a great place for conversation groups. Our friends are not particularly impressed with the fact that this is a solar house, but they seem to like the contemporary look, wood finishes, plants, and the warm feeling of the interior.

The solar aspects of the house are easy to deal with. A thermostat at the top of the north stair tower is set to activate the fan that charges the rock bed below the main lower floor. The floor is warm most of the time from heat stored in the rock bed. We have played with adjusting the single-point thermostat to different settings in the morning from 70 to 80°F (21–26.5°C) and it doesn't seem to make any dif-

Upper-level master bedroom affords views and privacy. Hauman residence.

Second-level living area opens to below. Hauman residence.

1. direct winter solar gain
2. air to rockbed
3. fan
4. rockbed
5. air from rockbed
6. radiant thermal storage mass
7. night air cooling ducts
8. energy-efficient fireplace
9. vent windows
10. roof vent turbine

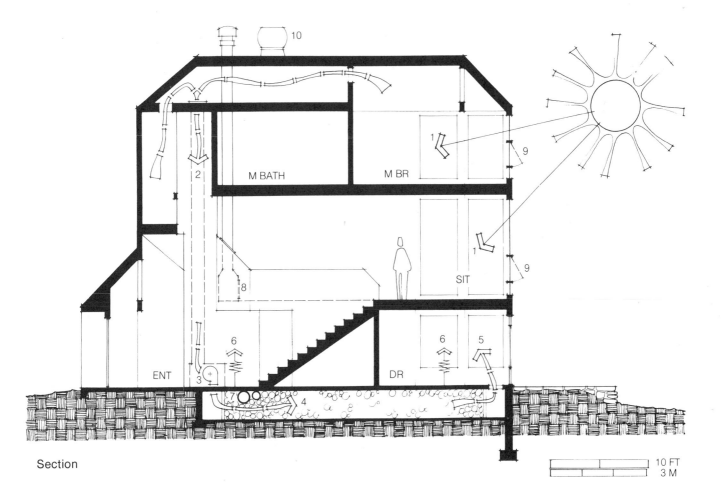

Section

10 FT
3 M

ference. When the room temperature at the top of the stairway hits the set temperature, the fan turns on and draws the sun-warmed stratified hot air down to the rock bed. During a clear winter day, the fan runs from 9 A.M. to 5 P.M. However, the fan and duct are inadequate. Temperatures in the top master bedroom have hit 106°F (41°C) with the fan circulating at full capacity. We are now having a larger fan and duct installed with a differential thermostat.

We have been pleasantly surprised at how easy it is to cool the house in summer. The stack effect of opening high and low windows causes a strong upward air motion that cools the house well. It often gets muggy in this area, and we are amazed that the house stays cool and dry.

Project Data Summary

Project Information

Project: Hauman residence
Guilford, Connecticut
Designer/builder: David Conger, Steven Conger, Paul Lytle
Guilford, Connecticut

Climate Data

Latitude	41°N
Elevation	23 FT (7m)
Heating degree days	5,880
Cooling degree days	365
Annual percent possible sunshine	60%
January percent possible sunshine	50%
January mean minimum outdoor air temperature	21°F (−6°C)
January mean maximum outdoor air temperature	38°F (3.5°C)
July mean minimum outdoor air temperature	61°F (16°C)
July mean maximum outdoor air temperature	80°F (26.5°C)
Climate features: harsh winters; warm, humid summers	

Building and System Data

Heated floor area	2,500 FT^2 (232m^2)
Solar glazing area	
Direct gain	532 FT^2 (49.5m^2)
Thermal storage heat capacity	
Tiled concrete floor	11,710 Btu/°F
Rock bed	14,060 Btu/°F

Performance Data

Building load factor	9.0 Btu/DAY°F FT^2
Auxiliary energy (heating)	44.9 MMBtu/YR
Auxiliary energy (cooling)	3.9 MMBtu/YR
Solar heating fraction	57%
Night ventilation cooling fraction	91%

SUNHOUSE COMPLEX: San Francisco, California

Mass housing is eminently suitable for passive solar designs, as these elegant solarium townhouses demonstrate. Sunhouse Complex, San Francisco, California.

Open three-story solarium is the focal point for each townhouse. Sunhouse Complex.

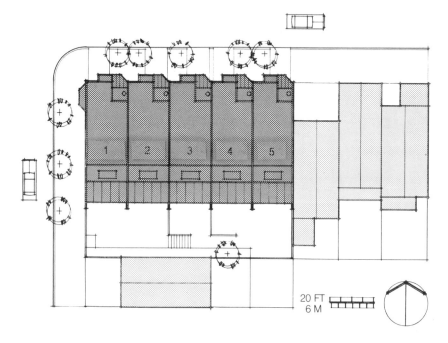

Site plan

Mass housing is often the only cost-effective dwelling option in urban and suburban communities, and the application of passive solar design principles to this important type of building has been slow. The bureaucratic stumbling blocks to implementation are numerous and inhibiting; building departments, government agencies, finance institutions, trade unions, and others are not responsive to innovation in this housing type. Also, the complexities of creating designs appropriate to the urban landscape are in themselves difficult. Coming up with a passive solar solution for this size project is doubly difficult. Consequently, the adventurous developer undertaking the challenge of building a passive solar housing project must be prepared to put forth a little more effort to succeed.

Zoe Works Architects of San Francisco worked with their client through the bureaucratic maze. The beauty of this complex is that it steps down a city hillside, gathering solar energy without compromising the public or private aspects of each family unit. The result is well worth the effort. This attractive complex boasts energy performance that may well become required standards for new buildings.

From the Architect

The Sunhouse project evolved from the outset as an opportunity to demonstrate the potential for passive solar space heating in an urban setting. The owner/developer's objectives were twofold: to develop a project that would be compatible with the family-oriented neighborhood, and to effectively utilize solar energy and energy-conscious designs. The project is located in the Western Addition Redevelopment Agency of San Francisco and, as such, had to meet the various requirements of that agency. Specifically, we had to gain neighborhood approval and win agency selection over several other developers who submitted more conventional proposals.

Our design philosophy evolved from the needs of urban dwellers—their life patterns and space requirements. This philosophy is summarized as "maximum solar utilization with minimum user involvement." Most urban dwellers are not available to manually monitor, adjust, and control solar systems. Generally, they leave in the morning, not returning until the evening. They take frequent vacations, ranging from long weekends to several weeks. Cognizant of these factors, we designed a system which would typically require no more effort to operate than setting the thermostat, yet could be manually adjusted if necessary.

In an urban environment, the conditions which restrict solar design, particularly passive solar design, are usually more acute than in rural or suburban locations. The factors of limited site, solar access, proximity and character of existing buildings, older and more restrictive building regulations, existing facilities, and high labor costs were critical design considerations for this urban project.

The hillside site, with its good solar exposure, was ideal. Our design ob-

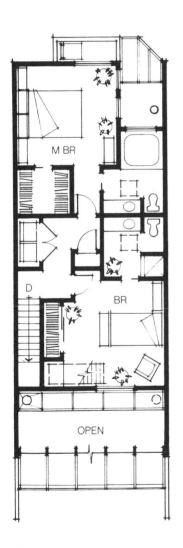

Upper level

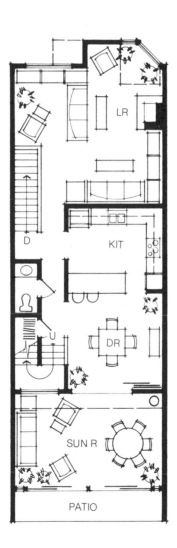

Main level

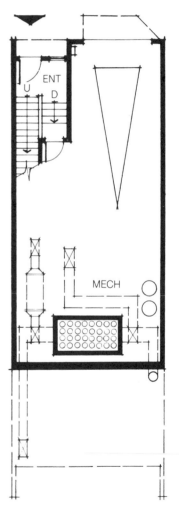

Entry level

10 FT
3 M

jectives were to create a building configuration and shape reflective of its use and functions, yet expressive of the architectural heritage of the Bay Area. By sloping the south-facing solarium glass wall and the side enclosure walls at sixty degrees, we emphasized the optimum winter sun angle. The bay windows, exterior wood shingles, and the interplay of building volumes on the north facade are to provide an interesting visual effect and to identify with the local architectural character.

The San Francisco Building Department's principal objective in regulating solar projects has been to insure that the equipment conformed to the city's well-regarded plumbing code. Since architectural components themselves, rather than conventional heating and cooling equipment, would supply the thermal needs of Sunhouse Complex, the city found its building code was inadequate to deal with a passive solar project. Engineering evidence proved that a passive design employing green-

houses or solariums with thermal storage could provide most of the heating needs and nearly all the cooling needs naturally.

However, the building inspector considered the greenhouse a habitable space, which violated the building code. We successfully argued that the greenhouse-enclosed patio was really a solarium, which the code permits to be used for living purposes. In addition, the southern rooms of the building open into the greenhouse, instead of directly to the exterior, as

Sunspaces epitomize indoor/outdoor living. Sunhouse Complex.

Detail of solar "super heater." Sunhouse Complex.

1. winter solar gain
2. warm air to thermal storage
3. solar super heater
4. fan to charge thermal storage
5. phase-change thermal storage bin
6. air from thermal storage
7. interior warm air to thermal storage
8. fan to heat unit
9. conventional heating unit
10. insulating shutter
11. exhaust vent
12. solar water heater collectors
13. skylights

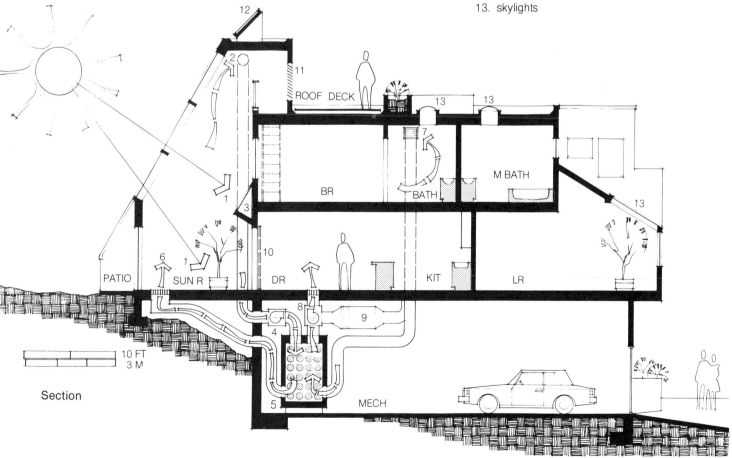

Section

the code's ventilation provisions require. Here, we showed that the two back rooms would receive adequate fresh air from the greenhouse, supplied directly through operable dampers in the glazing. The third code problem was the heat-storage unit of phase-change thermal material. The code contained no standards for such a storage arrangement. The phase-change thermal mass has a storage capacity adequate to supply the heating needs of each unit for four successive sunless days. However, the project owner had to agree to provide the townhouses with gas-fired or electric furnaces capable of handling the full heat load of each building, regardless of stored solar energy.

The final design centers around a large greenhouse in each townhouse unit. The attached greenhouse faces south at the rear of each unit and serves as solar heat collector, cooling ventilator, and usable indoor space. The sloping facade creates a large ground-to-roof enclosure to which the doors and windows of the two main levels open.

Solar radiation heats the walls and paving during the day. Excess heat is drawn by convection directly into the living areas, or by fans to the underground phase-change thermal storage, after first passing through a solar "super heater" (a collector area inside the greenhouse that increases the return-air temperature). At night, heat in the solarium's thermal mass reradiates directly to the space, while stored heat is drawn from the remote thermal storage bin to heat interior spaces on demand.

The system is operated by thermostatic controls. A differential thermostat switches on the hot-air intake fan when the temperature in the solarium reaches 10°F (5.5°C) higher than the temperature in the thermal storage unit, thus drawing air from the solarium to the storage area. In each unit, another thermostat switches on a second fan when the temperature falls to the comfort level set by the individual user, drawing warm air from the storage area to the interior. When the heat capacity in storage is depleted, the conventional heating unit automatically switches on, supplying the heating needs until no longer required or until the storage supply is replenished.

The summer cooling mode uses convection currents in the solarium to provide cool air to the living area. Again, outside air is continuously drawn into the solarium and the interior spaces through glazed openings. Hot air is then vented from the top of the solarium. This enables cool, fresh air to be continually drawn in, even on days with stagnant air.

From an Inhabitant

Living in Sunhouse produces some pleasurable surprises. Our initial discovery was that the solarium is very appealing at night. The glass provides a closed environment which allows for a pleasant moonlit patio setting on a cool evening, so very typical of San Francisco. We discovered that the three-story space allows for elegant entertaining, especially larger parties, for the scale of the room creates an open feeling. The tile floor provides a wonderful dance floor, and the heavy insulation is an extra bonus as a barrier against city noises. This is immediately apparent upon entering Sunhouse: you suddenly realize you are in an environment of complete peace and tranquility. This is further validated when you enter the solarium and have a private garden along a glass wall; the feeling it evokes is captivating. Daytime living is still the primary pleasure and, due to our irregular job hours, we are lucky enough to enjoy the house during the sunny hours. We often open the doors of the solarium and indulge ourselves in lounging around, reading, and acquiring a tan in delightful comfort.

The energy performance is great. We leave the thermostat at a decadent 70° (21.1°C) all year round. We use natural gas as a backup system for the central heat and hot water. These work automatically and with minimal expense. The gas portion of the utility bill is below the "lifeline" level created by our local utility for minimal users. This is true for all five owners in the development.

One of the more extraordinary experiences we have had repeatedly is observing the extreme caution people take when exposed to some of the innovative details in our house. It is as if they are looking at some new horseless carriage. Some people want to be reassured that the whole place won't blow up if someone pushes the wrong button. When visitors see all the ducts converging into the heat storage box, they usually approach the area as if it were radioactive. Even the water heaters are treated as if they have acquired a space-age status. Actually, it is fun to watch.

Project Data Summary

Project Information

Project: Sunhouse Complex
        San Francisco, California
Architect: Zoe Works—Garth Collier
        San Francisco, California
Builder/developer: Delameter Partnership
        San Francisco, California

Climate Data

Latitude	37.8 °N
Elevation	200 FT (61m)
Heating degree days	3,042
Cooling degree days	188
Annual percent possible sunshine	67%
January percent possible sunshine	56%
January mean minimum outdoor air temperature	46°F (7.5°C)
January mean maximum outdoor air temperature	56°F (13.5°C)
July mean minimum outdoor air temperature	53°F (11.5°C)
July mean maximum outdoor air temperature	64°F (17.5°C)

Climate features: mild, moderate, foggy winters; cool summers

Building and System Data

Heated floor area	1,840 FT² (171m²)
Solar glazing area	
Sunspace/per unit	460 FT² (42.5m²)
Thermal storage heat capacity	
Phase-change rods/per living unit (latent heat)	254,200 Btu

Performance Data

Building load factor	5.8 Btu/DAY°F FT²
Auxiliary energy (heating)/per unit	1.5 MMBtu/YR
Auxiliary energy (cooling)/per unit	3.2 MMBtu/YR
Solar heating fraction/average per unit	92%
Night ventilation cooling fraction/average per unit	69%

Vaulted roof, cantilevered deck, and integrated solar water heaters give a
unique shape to this redwood house. Aaron residence, Occidental, California.

A fine solar sculpture in the redwood forest. Aaron residence.

All criteria involved in designing a passive solar home must be weighed and reviewed carefully to ensure successful design. The architect may be very satisfied with a solution, but the broader judgment is made by society, whose evaluation is essential in the evolution of a new design. The architect, builder, and occupant should act in concert to evolve a new concept. Sometimes, as in this case, the individuals have very different reactions to a given solution. This difference should be noted and understood by all parties in order to refine the concept for future applications.

Architect Peter Calthorpe is a concerned and creative individual with a number of successful projects completed in the area of alternative energy architecture. Calthorpe and his partner, Sim Van der Ryn, apply passive solar principles in the planning and design of projects at all scales.

Calthorpe recommended this home for this book because "it is interesting from an architectural as well as a solar perspective. It is the conclusion, as it were, of the passive-discharge rock-bed concepts." This unique design is the result of a logical process attempting to reduce construction costs and improve performance by incorporating thermal storage mass with wood-frame construction. This hybrid concept works.

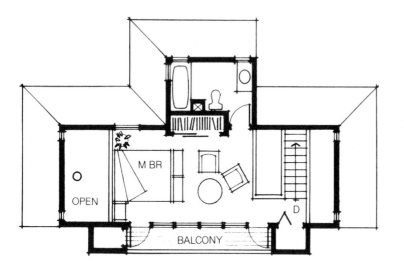

Loft level

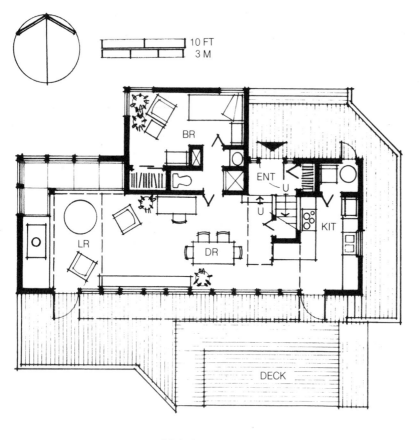

Main level

From the Architect

This small two-bedroom house employs a hybrid solar heating system. Like most of the construction in this area, this building is a two-story wood-frame structure on a continuous foundation. The basic rectangular shape, with the long side facing south, optimizes the building's surface area for solar gain. The site has good solar exposure, with an open meadow to the south; the north side is sheltered by a redwood grove. The mild climate allows freedom of window placement and two-story spaces. The overall building size is small, a conservation feature not to be overlooked.

The solar heating system uses a fan to transfer heat into the rock-bed thermal storage mass and allows natural radiation and convection to distribute the heat back to the living space. Because the system uses a fan to actively charge the storage while allowing for a natural or passive discharge, it is considered a hybrid. Specifically, dark-colored Venetian blinds are placed behind the double glazing on the south side to absorb the sunlight. A third layer of glass is placed inside the blinds to contain heat while a small fan draws air from the window cavity down into the rock bed. The heat from the window is deposited in a "vertically charged" horizontal rock bed below the floor. While dependent on a fan to operate, this collection system offers several distinct advantages. The glare, furniture fading, and large temperature swings associated with some direct-gain systems are eliminated by the louvers, while the views and light from the windows are maintained—the best of both worlds. On overcast days, the blinds can be raised to allow a direct heat gain from the diffuse light outside. At night, the blinds are closed, providing privacy and some insulation.

The rock bed is placed in the standard foundation crawl space. Perforated plastic drainpipes are spaced 16 inches (40.5cm) apart to create a return plenum with uniform air flow throughout the bottom of the rock bed. These pipes run into a larger manifold on the north side, which is connected to the fan. River rock is placed over the drain pipe, leaving a 6-inch (15cm) air space below the plywood floor to act as the intake plenum from the windows. The hot air from these windows enters the top plenum and distributes heat evenly over the entire surface of the rock bed. As the air moves downward through the rock, it deposits its heat in horizontal layers—the warmest layer at the top, the coolest at the bottom. The warm upper layer heats the floor above, which in turn radiates heat to the space. After depositing its heat, the air passes into the lower drainpipe plenum and is returned via a duct high on the second floor.

The passive heat-distribution system functions much like the old radiant floor systems. The vertical charging of the rock bed is critical to insure the even heating of the total radiant floor area. Yet, unlike the older hot-water-pipe radiant floors, this floor cannot be turned on and off. Therefore, the size of the system is designed to handle the average, rather than peak, heating load.

1. interior air to collector
2. single-layer interior glazing
3. dark-colored Venetian blinds
4. double glazing
5. warm air to rockbed
6. rockbed
7. insulation
8. perforated plastic pipe return plenum
9. return manifold to fan
10. radiant heat from floor

The upstairs bedroom has an open balcony overlooking a two-story space; heat rises naturally into this space. The first-floor bedroom has no radiant floor and does not receive heat directly from the solar system. Two ceiling fans prevent any extreme temperature stratification. A wood-burning stove provides backup heat.

Shading and ventilation are the primary cooling devices. In addition, the rock bed can be used to store night coolness. In this mode, cool night air is blown through the rock bed when outside temperatures fall below 60°F (15.5°C). During the next day, cool air is blown into the living space and the

warm interior air is circulated to the rock bed, to be exhausted the following night. The lower south windows are shaded in summer by the balcony and solar collectors above. The upper south windows use Venetian blinds to shade the interior. All spaces have cross-ventilation, and the fans can be used to create air movement during particularly still summer periods.

Although this building does not use purely passive features, its hybrid system offers some interesting alternatives for cases where southern views are desired but direct-gain systems are inappropriate. It offers control through operation of the blinds

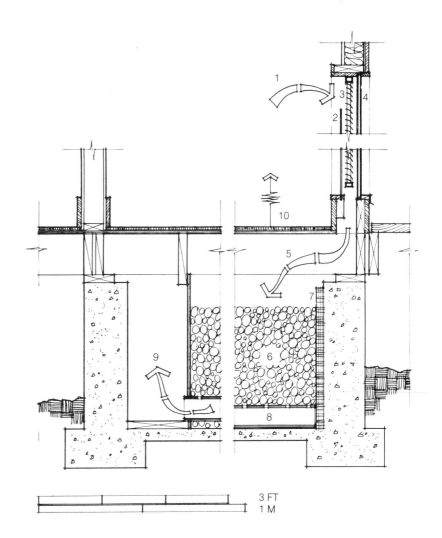

Collector/rockbed details

and floor registers. All of the system's components, including blinds, drain pipe, rocks, fans, and thermostat, are common building elements. Finally, and perhaps most significantly, the system is applicable to standard light-frame structures and requires no expensive masonry construction or additional interior space for heat storage mass or solar collection. Some disadvantages are that the system depends on electrical power to operate, is not as well known as other passive systems, and requires periodic cleaning of the windows and blinds.

After years of working with rock beds as a thermal-mass alternative for light-frame buildings, I have become wary. Although generally as cost-effective as masonry, they are more complex to design, construct and operate. My findings are that rock beds coupled with direct-gain spaces or greenhouses are problematic because of the low temperatures involved. Of all the rock-bed systems I have experimented with, the louver window is the most successful because of its inherent control and higher collection temperatures.

From the Inhabitant

The house has worked well. With the exception of a few nasty details, I am quite content. I find the technology simple enough to almost, but not quite, fully comprehend. It should be mandatory that owners are supplied simplified explanations and descriptions of their system, including what might go wrong, how to trouble-shoot, and a list of sources for help. Though I enjoy my low utility rates and feel I am contributing to important ecological matters, I also miss the security of not having qualified repair people available for questions and help. Not all solar-home owners are handy with tools or knowledgeable about systems!

On hot summer days, if there is a breeze, the house can be made fairly comfortable through cross-ventilation alone. If it is extremely hot, the fan helps, but it is noisy. Getting to the fan motor and oiling it is difficult because it is not easily accessible.

In the winter, the solar windows contribute warmth to the house. The versatile blinds offer two important benefits. In summer they act as sunlight reflectors, and in winter as solar absorbers. However, the blinds are clumsy to operate and difficult to clean. The iron stove provides most of the backup heat. Even in the coldest months, my total utility bill never exceeds $11 a month, $3 of which is for running the well pump. The solar water heater works quite effectively, too. From May to October, I simply turn off the electrical backup for the water heater.

I think were I to do it over, I would have a different design—simpler, more traditional—and I would choose a totally passive system. The overall design is so pleasing that I often feel as if I am living inside a fine sculpture.

The living room opens to the second floor, allowing heat to rise naturally. Aaron residence.

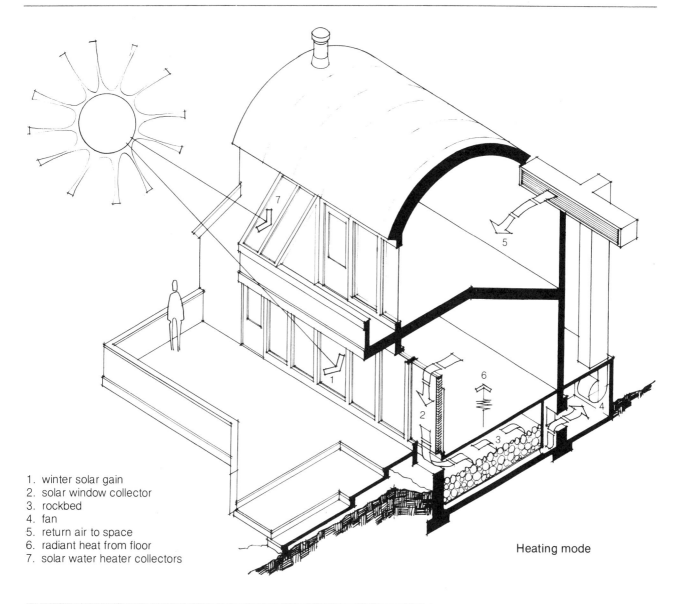

1. winter solar gain
2. solar window collector
3. rockbed
4. fan
5. return air to space
6. radiant heat from floor
7. solar water heater collectors

Heating mode

Darkened blinds between glass collect heat and control light. Aaron residence.

Project Data Summary

Project Information

Project: Aaron residence
Occidental, California
Architect: Peter Calthorpe
Inverness, California
Builder: Wayne Cartwright
Thailand

Climate Data

Latitude	38 °N
Elevation	300 FT (91.5m)
Heating degree days	3,019
Cooling degree days	37
Annual percent possible sunshine	78%
January percent possible sunshine	64%
January mean minimum outdoor air temperature	36°F (2°C)
January mean maximum outdoor air temperature	55°F (12.5°C)
July mean minimum outdoor air temperature	48°F (9°C)
July mean maximum outdoor air temperature	83°F (28.5°C)
Climate features: moderate, cloudy winters; hot summers	

Building and System Data

Heated floor area	1,111 FT^2 (103m^2)
Solar glazing area	
Louvered glass	172 FT^2 (16m^2)
Direct gain	104 FT^2 (9.5m^2)
Thermal storage heat capacity	
Rock bed	15,400 Btu/°F

Performance Data

Building load factor	9.1 Btu/DAY°F FT^2
Auxiliary energy (heating)	10.2 MMBtu/YR
Auxiliary energy (cooling)	5.4 MMBtu/YR
Solar heating fraction	55%
Night ventilation cooling fraction	80%

GREENSTEIN RESIDENCE: Woodland Hills, California

Turbines, a great flap, and rolling canvas shades earmark a distinctive solar
dwelling. Greenstein residence, Woodland Hills, California.

The most popular question posed by clients, builders, and developers is, "How much does a passive solar house cost?" The standard reply by knowledgeable architects and designers is that it can, and indeed should, be built for the same cost as a high-quality, well-built conventional home—perhaps even less in states where solar- or energy-conservation tax credits apply.

Architect Rob Wellington Quigley created this Southern California house/sculpture with simple solar design features. By carefully rearranging traditional architectural elements, using typical construction details, employing standard materials, and creatively adapting existing hardware to special uses, he came up with this hybrid design that cost about the same to build as a conventional home.

It also heats and cools itself with little auxiliary energy and automatically adjusts its interior climate to the weather.

A standard forced-air furnace is used for auxiliary heating. The fan of this unit is used to draw warm air from an isolated solar water wall to heat the interior indirectly, or to provide preheated air to the furnace. During hot spells, air is circulated through a night-cooled subfloor rock bed for cooling. These features are packaged as a dynamic, playful experience which points to one direction of a new, rational architecture.

From the Architect

The Greenstein house was designed for a young couple with one child. The modest-size structure uses standard two-by-four stud framing and catalog parts throughout to meet a tight budget. The house is organized around a high belvedere tower pierced by a linear gallery. Its gallery begins at the garage as the entry bridge, passes under a window seat and through the front door to reveal the dining room below at the left, the living room on the right, and the library above. The gallery then cascades through the base of the belvedere, past the kitchen and living deck, to the two bedrooms beyond. A centrally located kitchen allows supervision of the child's room and southern play yard. The formal result of the structure is as much a response to the hot climate as to the clients' personal style—organized but flexible.

Exterior materials, while simple, vary to respond to the conditions of the site and microclimate. The cooler north and east surfaces of the building are sheathed in dark-covered wood to absorb radiation, while the warmer south and west surfaces have silver-colored asphalt shingles to reflect excess radiation. Thus, the building's surfaces, as affected by sunlight, result in a sort of environmental graphic.

Looking up past the movable flap to the turbine-topped belvedere. Greenstein residence.

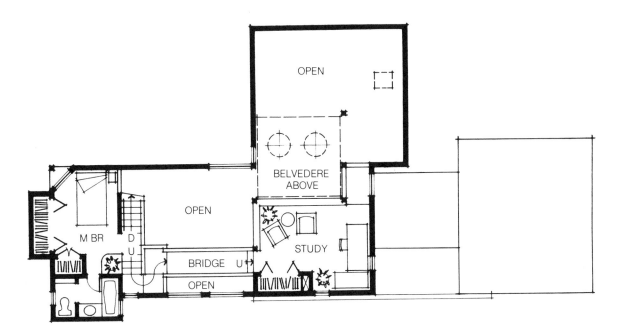

Loft level

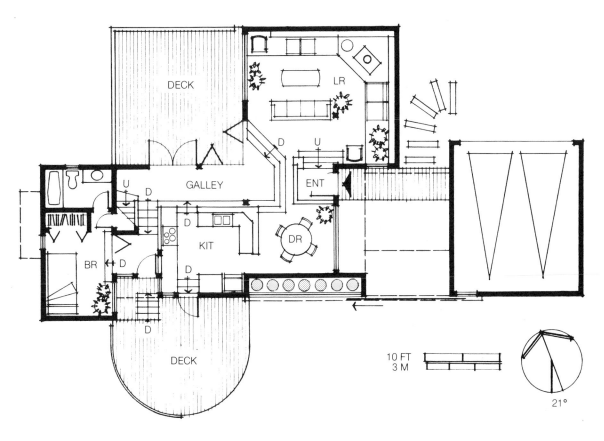

Main level

10 FT
3 M

21°

The wood windows frame selected views and are carefully shaded by landscaping on the hot west side and by sunshades on the south. The canvas sunshades wind around sprockets at the corners of the windows. They are designed to allow only the warming winter rays to penetrate the house.

The house has two separate space-conditioning systems. Solar space heating is provided by a single-glazed water wall of black drums in a "solar box." A standard forced-air furnace is used as the auxiliary. Return air is routed into the south-facing solar box to be preheated before it reaches the furnace. The furnace ignites only if air from the solar storage is not warm enough. At night, an insulated barn door is closed over the south glazing to minimize heat loss. The second system provides cooling. A large rock bed is contained in the crawl space under the child's bedroom. A fan on a timer blows cool night air through the rock bed during hot spells.

The open spaces organized around the tower allow the house to act as a chimney: hot air rises to the belvedere and naturally vents through large turbines above. When it is inconvenient to allow this natural flow of air, a large flap or damper, like a kinetic sculpture at the throat of the tower, is used to close off the belvedere volume. This "variable volume space" reduces the amount of interior air to be heated or cooled by the mechanical system. On days too hot for natural ventilation, the windows and damper are closed, and the house air is routed through the cold rock bed and distributed through the same duct system used for heating. There is no backup for the cooling.

Domestic water is heated by a standard active solar system. The collector is located on the belvedere, just below the large roof turbines. The combination of these two functional elements creates an energy sculpture that dominates the building's form.

From the Inhabitant

We have only leased this house for four months—through the hottest part of the summer, the cool nights of fall, and what passes for winter in Southern California. Yet this short period of time has provided new experiences and insight into the house and solar energy. It has demonstrated that it is important to achieve a house which is energy- and resource-conscious, not merely an energy-conserving house. The neat diagrams of the energy flows emphasize the central belvedere tower, around which all rooms are placed. However, the tower was not a preconceived approach to energy flow, but was a result of a question that the architect asked the owners. "What would you like to live in?" In their response—"a lighthouse"—he found a spatial concept which he intuitively saw as a means to move heated air through the series of open and interrelated spaces.

Our summer routine is simple. We use a series of defense strategies. Each morning, I move around the house to ensure that all windows are closed tight—the first line of defense. At midday, as the temperature rises in the house, we open the dampers below the roof turbines, already lazily turning in the first expected breeze, releasing some built-up warm air. Later, if the temperature continues to rise, we close the dampers and swing the great rotating translucent flap across the tower, reducing the volume of air to be cooled. Then we turn on the furnace fan, which draws cool air from the rock bed beneath the children's room.

On very hot days, we wait for that special time in the afternoon when the ocean breezes come up from the southwest. Frequently, the interior temperature has reached a peak of 85°F (29.5°C) despite all defense strategies, but we do not want to open the doors while it is still warmer outside than within. We liken this to the

Mechanical schematic

1. hot-air return
2. conventional forced-air vent
3. to house outlet registers
4. cool night air intake
5. rockbed fan
6. rockbed
7. water wall
8. interior air return
9. warm night air exhaust

The home's playful, open, and delightful interior; the flap can be seen in the
foreground. Greenstein residence.

timing required for removing a soufflé from the oven—neither too soon nor too late. To inform ourselves of that time, we moved a wind chime to the south deck. It begins to ring tentatively with the first breezes, then ever more furiously as the breeze picks up. At that point, we throw open both the south and north deck doors and the front door; the cross-ventilation instantly sweeps through the house, creating a breeze. We close the doors in early evening and leave the windows open through the night to cool the house for the next day's cycle.

During the heating period, the house is buttoned up during the evening and the turbine dampers remain closed. The tower flap is usually left open, although it would be more effective if it, too, were closed. The source of winter heat would normally be the water wall. However, we sometimes have difficulty remembering to open the insulating exterior door in the morning and to close it at night. This continues to be a problem even though the wall is designed with a built-in reminder: when the wall is open, it slides across an overhead track suspended from the wall of the house and the adjacent wall of the garage, thus closing off the entry court that lies between them. Usually, upon returning home, we realize we forgot to open it. This is a modest problem and simply one of developing mildly different habits.

These are small issues, however, hardly worth mentioning other than as a record of how it feels to live in a solar house and to be in tune with the daily and seasonal climate. As I sit writing this, with the morning sun streaming in every window and playing shadows and spots of light on the wall, I can only smile gratefully for whatever "inconveniences" the solar house might create, while being thankful for the living experiences that this piece of architecture has provided.

1. solar gain to water wall
2. warm air to thermal storage
3. conventional forced-air vent
4. return air from thermal storage
5. pivoting isolation flap (closed)
6. solar water heater collectors

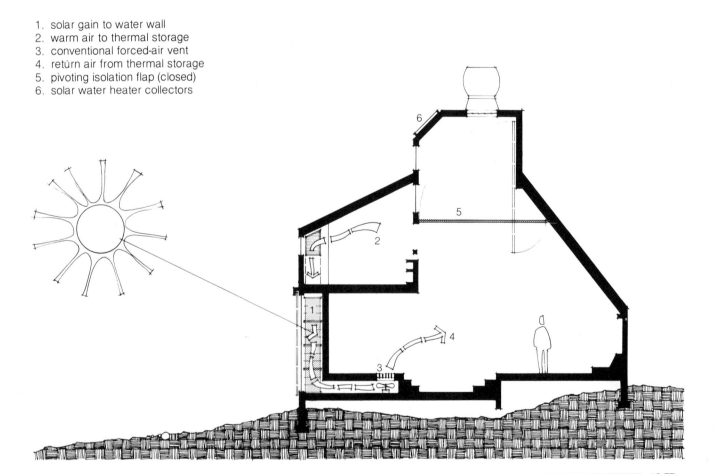

10 FT
3 M

Heating mode

Canvas sunshades protect the house from the afternoon sun. Greenstein residence.

1. cool night air intake
2. rockbed fan
3. warm night air exhaust
4. conventional forced-air vent
5. cool-air supply
6. pivoting isolation flap (open)
7. roof turbine exhaust
8. rolling canvas shades
9. sliding insulating door
10. solar water heater collectors

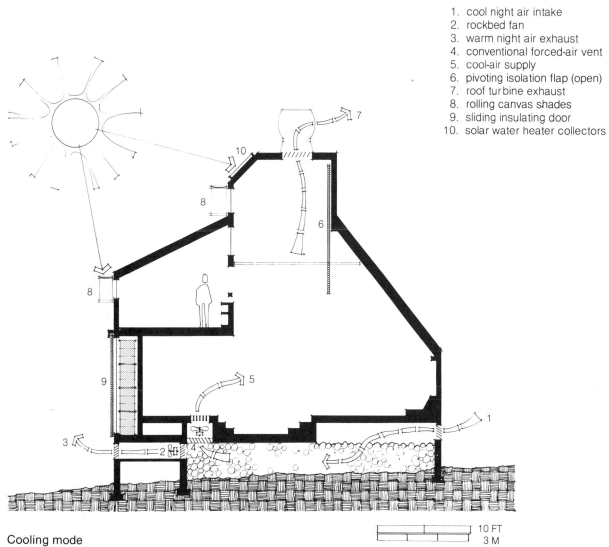

Cooling mode

10 FT
3 M

Project Data Summary

Project Information

Project: Greenstein residence
          Woodland Hills, California
Architect: Rob Wellington Quigley, AIA
            San Diego, California
Builders: Scott Carde and Wade Killefer

Climate Data

Latitude	34.2 °N
Elevation	1,010 FT (308m)
Heating degree days	1,245
Cooling degree days	1,185
Annual percent possible sunshine	73%
January percent possible sunshine	69%
January mean minimum outdoor air temperature	42°F (5.5°C)
January mean maximum outdoor air temperature	62°F (16.5°C)
July mean minimum outdoor air temperature	56°F (13.5°C)
July mean maximum outdoor air temperature	82°F (27.5°C)
Climate features: hot, dry summers; wild winters	

Building and System Data

Heated floor area	1,600 FT^2 (148.5m^2)
Solar glazing area	
Direct gain	89 FT^2 (8m^2)
Water-drum glazing	183 FT^2 (17m^2)
Thermal storage heat capacity	
Rock bed	8,040 Btu/°F
Water drums	8,259 Btu/°F

Performance Data

Building load factor	8.8 Btu/DAY°F FT^2
Auxiliary energy (heating)	1.9 MMBtu/YR
Auxiliary energy (cooling)	9.1 MMBtu/YR
Solar heating fraction	87%
Night ventilation cooling fraction	72%

AUTONOMOUS DWELLING VEHICLE: St. Louis, Missouri

On the edge of technology, this transportable habitat may be part of the future.
Autonomous Dwelling Vehicle, St. Louis, Missouri.

High-tech-inspired designers Ted Bakewell III and Michael Jantzen like to create designs which operate on the leading edge of technology. This Autonomous Dwelling Vehicle was conceptualized and built by Bakewell, Jantzen, and Ellen Jantzen. Their goal was to create a self-sufficient, transportable habitat that incorporates much of the state-of-the-art life-support hardware available. These elements, which cost less than $20,000 for materials, are incorporated into a slick, efficient, attractive package that serves as Ted's residence.

The Bakewell-Jantzen team has dealt candidly with a housing need of both present and future importance, which many architects and designers do not consider urgent. The Autonomous Dwelling Vehicle is a hybrid design and uses more advanced technology than most solar houses. It also uses on-site energy and resources to maintain most life-support systems, including space conditioning, refrigeration, and waste recycling.

Cooking presently requires off-site renewable fuel in the form of alcohol. With a little more menu selectivity and cooking-system efficiency, Bakewell could survive in his environment needing only supplemental food and transportation assistance.

This spacy earth capsule may well be a miniversion of future human architecture and life-support systems. For the present, it is a lovely expression of high-tech functionalist art.

From the Designers

Design features:

1. Transportable on existing highways, local roads, rain roads, river barges, and ocean freighters within all U.S. limitations on height, width, and weight
2. Adaptable to water floatation
3. Suitable for helicopter airlift
4. Capable of being set up on uneven terrain
5. Stable in high winds in a full tie-down
6. Comfortable for two people in a variety of climates with adequate storage and psychological spaciousness
7. Independent from all external power sources and from non-renewable fuels
8. Autonomous from all water and sewer-main hookups, without dependence on chemical additives or nonecological disposal
9. Mass-producible with currently available components and technologies; cost-competitive with comparably sized luxury travel trailers
10. Lightning protected

Systems to achieve these advantages:

Chassis: Specially reinforced, commercially available mobile office frame

Exterior shell: Aluminized steel agricultural silo components, insulated with 3.5-inch sprayed polyurethane foam and fire protected

Heating: Two photovoltaic-energized, fan-assisted air collectors coupled with phase-change thermal storage mass; air-lock/greenhouse and controllable skylights provide additional solar gain; wood/wastepaper–burning backup

Electricity: Four photovoltaic panels with reflectors and wind generator; twelve-volt battery storage; high-consumption conveniences such as kitchen blender and rotating-shaft power tools are pneumatic, run off of a

Wind and sun power this very mobile home. Autonomous Dwelling Vehicle.

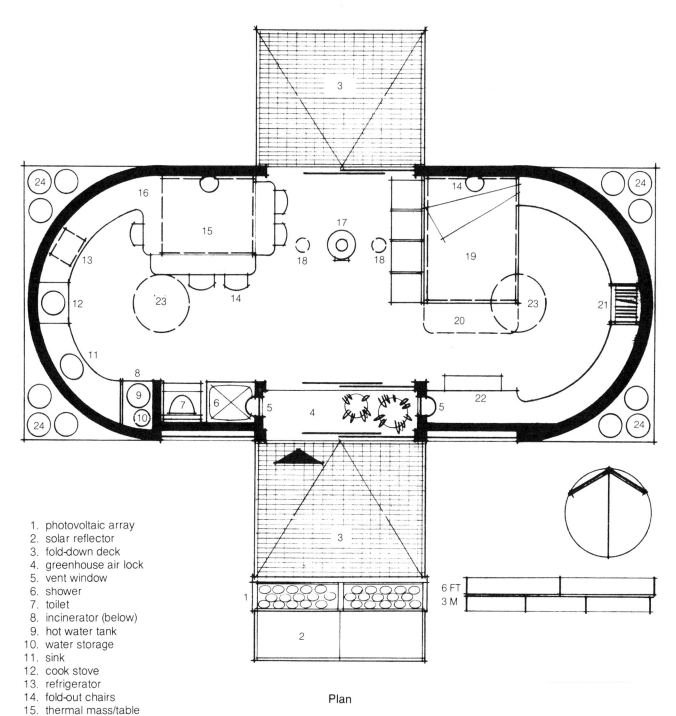

Plan

1. photovoltaic array
2. solar reflector
3. fold-down deck
4. greenhouse air lock
5. vent window
6. shower
7. toilet
8. incinerator (below)
9. hot water tank
10. water storage
11. sink
12. cook stove
13. refrigerator
14. fold-out chairs
15. thermal mass/table
16. recycling bins
17. energy-efficient wood stove
18. roof turbine vents
19. thermal mass/bed
20. fold-out extension for sleeping
21. closet
22. bench/storage/study
23. skylights
24. outside storage bins

6 FT
3 M

universal air motor; compressed air delivered by high-volume foot pump and wind generator

Lighting: Four high-frequency, dimmable fluorescent fixtures (developed for U.S. space program), three aircraft reading lights

Refrigeration: Six-and-a-half-cubic-foot (0.18m³) superinsulated box with solid state cooling; winter operation relies on through-the-floor heat exchanger utilizing ambient outside temperatures; unique door design increases efficiency

Cooking: Two denatured-alcohol burners, scrap wood/wastepaper stove, solar concentrator cooker, solar food drier designed into the skylights

Water: Rain collection system with on-board four-stage purification and used water recycling.

Water heating: Active solar collector using heat exchange fluid and photovoltaic powered pump; wood/wastepaper incinerator built into water storage tank for backup

Waste: Human and organic kitchen wastes composted in a waterless, chemical-free toilet

Recyclable material: Presorted in detachable bins built into the kitchen counter top

Storage: Multiple storage utilizing a variety of wall and ceiling pockets, industrial stacking trays, under-floor compartments

Comfortable and relatively stable interior temperatures were easily achieved with minimal use of the wood stove. However, during periods of excessive cloudiness and subfreezing temperatures with high winds, the wood stove must be fed at two- to four-hour intervals to maintain comfort. Thermal performance in cold, cloudy weather is significantly reduced when skylight and glass-door insulation are removed for natural light and viewing during the day. In periods of very cold but sunny weather, stable 68–78°F (20–25.5°C) interior temperatures are easily achievable.

Presently, two or three consecutive clear, sunny days are required to change the phase of the thermal storage mass. This will be reduced by balancing the collector air-flow rate and by improving the edge seal on the reverse-flow dampers.

The electrical storage capacity has not been fully tested or exploited to date. Intermittent sunshine has proved sufficient to power frugal use of lighting, television, stereo, and solar collector fans. Domestic solar water pump has been inactive most of the winter due to a failure of the air-bleed valve. Electricity for winter operation of the refrigerator has not been necessary. A ten-day sunless period in December-January demanded frugal electrical use to maintain primary functions. This prompted the addition of eighty watts of photovoltaic capacity, and a wind generator.

Water collection exceeds expectations even with low rainfall. A hard ten-minute rain easily fills the fifty-five-gallon storage bladder. The used-water recycling efficiency is improved if the water is cycled through the filter system immediately after use to avoid putrefaction in the holding tank.

There was excessive heat loss through the toilet vent, incinerator stack, and wood-stove chimney. We have recently installed an air-to-air heat exchanger at the toilet vent, an outside combustion air intake, and an airtight firebox door at the incinerator.

Future modifications to improve performance include the following:

1. Install insulated interior jacket at wood-stove chimney pipe.
2. Replace deficient edge seals on the insulating shades at glazing.
3. Install improved vapor barrier below floor.
4. Provide carpeting over tile flooring during winter.
5. Repair wind-skirt zippers around support chassis.
6. Improve gray-water filter system.
7. Insulate toilet vent stack to reduce interior condensation.

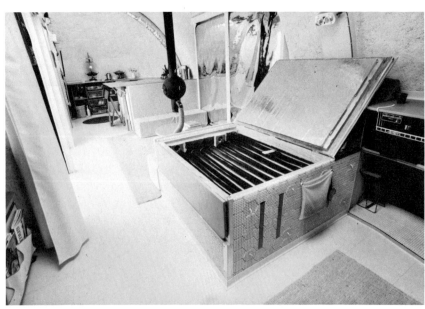

Thermal storage under the bed. Autonomous Dwelling Vehicle.

The kitchen is compact yet very complete. Autonomous Dwelling Vehicle.

1. solar air collector
2. phase-change thermal storage
3. photovoltaic array
4. solar reflector
5. fold-down deck
6. pneumatic closure skirt
7. skylight
8. wheel assembly and chassis

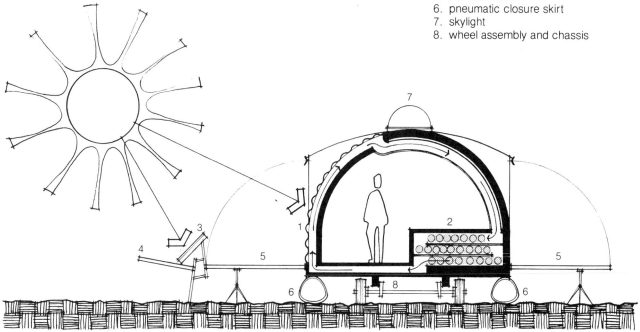

Section

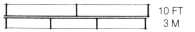

10 FT
3 M

Project Data Summary

Project Information

Project: Autonomous Dwelling Vehicle
            St. Louis, Missouri
Designers: Michael Jantzen, Ted Bakewell
            St. Louis, Missouri
Builders: Ted Bakewell, Michael Jantzen, Ellen Jantzen

Climate Data

Latitude	38.8 °N
Elevation	535 FT (163m)
Heating degree days	4,750
Cooling degree days	1,475
Annual percent possible sunshine	59%
January percent possible sunshine	52%
January mean minimum outdoor air temperature	23°F (−5°C)
January mean maximum outdoor air temperature	40°F (4.5°C)
July mean minimum outdoor air temperature	69°F (20.5°C)
July mean maximum outdoor air temperature	88°F (31°C)
Climate features: cold winters; warm summers	

Building and System Data

Heated floor area	400 FT² (37m²)
Solar glazing area	
Greenhouse	101 FT² (9.5m²)
Convectors	132 FT² (12m²)
Photovoltaic panel	80 peak watts
Thermal storage heat capacity	
Phase-change thermal storage rods (latent heat)	300,000 Btu

Chapter 7

INTO THE FUTURE

A mistake is often made in predicting too far and too much. We do not intend here to predict; rather, we aim to illustrate and explain the trend in passive solar architecture. This new growth will blossom with many subtle branches. It will grow, divide, and serve the designers, builders, and home owners logically and beautifully, continuing the long history of architectural growth.

Energy cannot, as far as we know, be created or destroyed. Thermal energy, as it is used in most buildings today, is misused and wasted. The development of passive solar architecture makes use both of past knowledge and experience and modern materials and technology to create more efficient and comfortable buildings. Today's passive solar prototypes and system generics are forerunners of a new kind of energy-conserving building—a crude glimpse at the shape of things to come.

The design responsibility which we accept from this day forward is committed to better use of energy. Whether to power a space craft, energize a life-support system, or run an electric razor, a frugal energy ethic must prevail. With the next generations of architectural evolution, we will create dwellings that will more efficiently interact with energy and earth.

It is not difficult to speculate about the future of passive solar architecture: the ultimate design will create autonomous dwellings that can fully energize themselves anyplace in the world. Through our modern understanding of physics and ecology, we are on the verge of achieving this ability. The materials and systems that will be used are yet to surface. The buildings reviewed in this book are a beginning, and they also represent a significant shift from the standard concepts of historical and twentieth-century building practice. Sophisticated thermal storage methods, on-site electrical generation techniques, and microprocessor control mechanisms are only a few of the promising design options available today and already a part of contemporary design.

Wherever man settles, architecture will most likely produce radically different structures from the ones we know. Solar energized and conditioned buildings are a logical next step in the development of energy independence. The disassociation of buildings from the umbilical cord of central power generation could in itself significantly alter the way our buildings work, look, and feel. Photovoltaic surfaces integrated with the building's exterior are here today. Interior walls, ceilings, and floors could accept or emit thermal energy to maintain constant comfort levels with instant responsiveness to user demand. Soundless space-conditioning systems will require only enough air motion, filtration, and exchange to assure comfort. No longer will we put up with hot and cold spots, or dust-throwing forced-air systems.

We have begun to consider the options. New designs are already demonstrating more logic and efficiency than those of the past. However, logic and efficiency do not always satisfy the ideal of beauty. Both logic and beauty should be inherent in the energized architecture of the future. Mankind has the innate desire to create esthetically pleasing things. If we are forced by necessity to evolve longer-lasting, energy-efficient, high-performance buildings in order to survive, we will also be required to include quality as an integral element. The designers of this quality architecture will bear the responsibility of creating esthetic solutions that will satisfy the sense of beauty for decades to come. In the past, this has been optional; some day, it will be required.

APPENDICES

Appendix A: CALPAS3: Passive Solar Performance Simulation

The computer and computer program used in this book belong to the Berkeley Solar Group, an energy computer service and engineering firm. The computer is a Data General MV/8000 connected to GTE's worldwide data network, TELENET, accessible to 130 U.S. cities and 15 foreign metropolitan areas. The program CALPAS3 was originally developed at California Polytechnic State University by Professor Philip Niles for use in the *Passive Solar Handbook for California*. Microcomputer time-sharing interface is available. By using one's own terminal and a long-distance telephone link-up, rapid file transfer can be achieved.

To evaluate a passive home, the program first sets up a numerical picture of the home and its climate by reading building information provided by the architect and weather information on file in the computer. The building information consists of 30 to 100 word-and-number pairs that describe the windows, walls, heat-storing objects, and so on. The weather information, originally obtained from the National Oceanic and Atmospheric Administration, is a year's worth of wind, temperature, sun, and other observations made every hour for more than 250 locations in the United States. Using this information, the program calculates heat flows and temperatures in the house for each hour of that year.

CALPAS3 is an hour-by-hour thermal network simulation program that uses ASHRAE heat transfer and ventilation algorithms; backward (implicit) differencing equations for modeling transient heat conduction in mass elements (eliminating stability problems and allowing consistent use of a one-hour time step); and combined radiant-convective coefficients, allowing very fast execution with little loss of accuracy.

The output reports show the amount of heating provided by the sun, the amount of heat gained or lost through conduction and infiltration, the amount of natural ventilation cooling, the amount of backup heating or cooling needed, and other important information. These values are reported for the hours, days, and months specified. A summary report shows the estimated total annual utility bill for backup heating and cooling.

Architectural features the program can evaluate include an attached greenhouse/sunspace, thermal envelopes, Trombe walls, mass walls, rock beds, floor slabs, and water walls. Also, walls and windows which face in any direction and have any tilt or amount of insulation, window overhangs, movable insulation and shutters, natural and forced ventilation, and evaporative cooling can be modeled.

A user manual supplied with the service helps in selecting a computer terminal and modem for telephone transfer. A CALPAS3 manual is also provided to help the user understand how to set up the building input file and how to interpret the output reports.

Modeling Procedure

Site factors such as summer shading of window, wall, and roof surfaces by trees were generally not accounted for in the analyses because of lack of sufficient data. Additionally, in all the computer runs, the temperature in both the passive house and nonsolar (reference) house was maintained between 65°F (18.5°C) and 80°F (26.5°C) for comparison purposes. In this way, the auxiliary energy used by both passive and reference designs can be used to calculate heating and cooling fractions. In reality, the house temperature might be allowed to swing lower than 65°F (18.5°C) and higher than 80°F (26.5°C) due to the mean radiant temperature of the interior, as well as air motion, clothing, or use patterns.

In order to determine solar heating and night ventilation cooling fractions, a solar building must be compared to a similar nonsolar building without passive features. The nonsolar reference building is identical to the solar building except for the following:

1. South glazing is reduced to 7 percent of the heated floor area.
2. Entire floor slab is carpeted.
3. Retaining walls are replaced by insulated frame wall and no berm is provided.
4. Interior mass walls, rock beds, and other thermal storage units are removed.
5. No movable window insulation is provided.
6. No summer shading is provided for glazed areas.

Thus, the reference building performance represents the operation of a conventional nonsolar building, although the insulation values and double-glazing factors are the same as the solar house's, and may be higher than a conventional building.

As predicted by CALPAS3, the amount of auxiliary energy required for heating the buildings modeled in this book are generally within 10 percent to 15 percent of the actual recorded energy consumption for those projects which supplied comparable recorded data. This is quite admirable considering the potential differences between the simulation assumptions and the actual building conditions. For example, consistent control of movable insulation, shading, and outside air exchange rates are difficult to achieve in practice, whereas CALPAS3 controls these factors accurately.

The predicted amount of auxiliary energy required for cooling is generally higher than that required by the occupants of these buildings. One reason is that during summer months, partial shading of the building exterior, especially roof surfaces, can have a significant effect on the radiation absorbed and, subsequently, the heat conducted into the building.

Another major factor influencing the comparison between predicted and actual performance is weather data. The weather tapes used for CALPAS3 represent long-term average conditions for a location, whereas the weather conditions existing during the measurement interval may differ substantially from the norm. Although the results of the computer simulations closely predict auxiliary heating and cooling requirements, it must be understood that, as a result of site, climate, and other factors, the actual performance provides the most accurate means of evaluation.

Appendix B: Conversion Factors

Unit	x Conversion Factor	= Converted Unit
Length		
in	2.54	cm
m	39.37	in
cm	0.394	in
Area		
cm^2	0.155	in^2
in^2	0.000645	m^2
ft^2	0.092903	m^2
mi^2	2.58999	km^2
km^2	0.386	mi^2
in^2	6.451	cm^2
Volume		
in^3	0.0016	cm^3
cm^3	0.061	in^3
ft^3	0.028	m^3
gal	3.78544	l
ft^3	748	gal
gal/ft^3	0.02454	l/m^3
yd^3	0.7645	m^3
Weight		
lb	454	g
lb	0.454	kg
kg*	2.205	lb
Energy		
ft-lb	1.29×10^{-3}	Btu
Btu	778	ft-lb
Kcal	3.968	Btu
cal	4.186	joule
joule*	9.478×10^{-4}	Btu
Btu	1055.1	joule
kwh	3413	Btu
Btu	2.93×10^{-4}	kwh
hp	2544	Btu/hr
hp	745.7	watt

Unit	x Conversion Factor	= Converted Unit
Heat Flux		
$cal/(cm^2)(sec)$	13.272	$Btu/(hr)(ft^2)$
W/cm^2	3171	$Btu/(hr)(ft^2)$
$Btu/(hr)(ft^2)$	3.154×10^{-4}	W/cm^2
$Btu/(hr)(ft^2)$	3.154	W/m^2
langley	1.00	cal/cm^2
langley	3.687	Btu/ft^2
ly/min	221.2	$Btu/(hr)(ft^2)$
Btu/hr	0.293	watt
watt*	3.413	Btu/hr
ton air conditioning	12,000	Btu/hr
Temperature		
°F	+ 459.7	°R
°R	− 459.7	°F
°C	+ 273.1	°K
°K*	− 273.1	°C
°C	= 5/9(°F − 32)	
°F	= (9/5°C) + 32	

Abbreviations and Symbols

inch	in
foot	ft
yard	yd
centimeter	cm
gallon	gal
quart	qt
pint	pt
liter	l
meter	m
pound	lb
gram	g
kilogram	kg
foot-pound	ft-lb
British thermal unit	Btu
calorie	c
kilocalorie	kcal
joule	J
kilowatt hour	kw-hr
horsepower	hp
Watt	w
langley	ly
Fahrenheit degree	°F
Celsius degree	°C
Rankine degree	°R
Kelvin degree	°K

* Units followed by an asterisk are International Standard (SI) units.

Appendix C: Floor-Plan Nomenclature

BATH	Bathroom
BKFST	Breakfast room
BR	Bedroom
D	Down
DR	Dining room
ENT	Entry
FP	Fireplace
FAM R	Family room
G BR	Guest bedroom
GRNHSE	Greenhouse
KIT	Kitchen
LAU	Laundry
MECH	Mechanical room
M BATH	Master bathroom
M BR	Master bedroom
PAN	Pantry
REC	Recreation room
SEW	Sewing room
SIT	Sitting room
SUN R	Sun room
STOR	Storage
U	Up

Appendix D: Glossary

Absorptance: The ratio of radiation absorbed by a surface to the total energy falling on it; expressed as a decimal fraction or a percentage.

Active solar energy system: A system that requires outside energy sources (for example, electricity) in conjunction with solar sources and techniques.

Adobe: A sun-dried, unburned brick of clay (earth), sand, and straw used in construction; sometimes stabilized with oil-based additives. Within the United States, adobe is used primarily in the Southwest.

Ambient temperature: Dry-bulb temperature of the air in the immediate microclimate, as in the average air temperature outside a building.

Angle of incidence: The angle that the sun's rays make with a line perpendicular to a surface. The cosine of the angle of incidence determines the fraction of incident direct sunshine that is intercepted by a surface.

ASHRAE: Abbreviation for the American Society of Heating, Refrigerating and Air Conditioning Engineers, Inc., 345 E. 47th Street, New York, New York 10017.

AS/ISES: Abbreviation for the American Section of the International Solar Energy Society. U.S. Highway 190 West, Killeen, Texas 76541.

Auxiliary system: A supplementary unit to provide heating or cooling when the solar unit cannot do so. This is usually necessary during periods of cloudiness or intense cold.

Azimuth: The angular distance, measured in degrees, between true south and the point on the horizon directly below the sun's position.

Back flow: The unintentional reversal of fluid flow in a distribution system. Also referred to as back siphoning.

Backup system: see Auxiliary system.

Barrel wall: see Drum wall.

Berm: A man-made mound or small hill of earth.

British thermal unit (Btu): A measure of heat; specifically, the amount of heat required to raise the temperature of one pound of water 1°F. One Btu is approximately equal to the amount of heat given off by burning one kitchen match.

Building load factor: An indication of heat loss through the building skin. As a rule of thumb, 3–6 is considered excellent, 6–9 is very good, and 9–12 average. Units are Btu/sq. ft.°F day.

Calorie: A unit of heat; the amount of energy needed to raise the temperature of one gram of water 1°C. One hundred calories is approximately equal to four Btus.

Celsius (°C): A temperature scale in which the freezing point of water is set at 0°C and the boiling point of water is set at 100°C.

Celsius temperatures can be derived from Fahrenheit temperatures by the equation $C = 5/9(°F - 32)$.

Chimney effect: The tendency of air to rise when heated, because of its lowered density. This principle is used to help cool a building by allowing hot air to rise and flow out through upper-level windows. This creates a subatmosphere which draws cooler outdoor air in through windows at a lower level.

Clerestory: A window placed vertically (or nearly vertical) in a wall above one's line of vision to provide natural light or ventilation.

Climate: The meteorological conditions, including sunlight, temperature, precipitation, humidity, wind, and weather patterns, that characteristically prevail in a particular region.

Coefficient of heat transfer: see U value.

Collection surface: The part of the solar collector where solar energy is collected, generally darkened to increase absorption. It is often called an absorber.

Collector, flat-plate: An assembly containing a panel of metal or other suitable material, usually with a dark color on the sun side, that absorbs solar radiation and converts it into heat. The panel is generally contained in an insulated box, covered with glass or plastic on the sun side to retard heat loss. Heat in the collector transfers to a circulating fluid, such as air, water, oil, or antifreeze, and then flows to where it is used immediately or stored for later use.

Collector, focusing: Also called concentrating collectors, these collectors use one or more reflecting surfaces to concentrate sunlight onto a small absorber surface.

Collector, solar: A device used to collect solar energy and convert it to heat.

Collector tilt: The angle between the inclined surface of a solar collector and the horizontal plane. A collector surface receives the greatest possible amount of direct sunshine when its tilt is perpendicular to the sun's rays.

Condensation: The process by which water is released from the air.

Conductance (C): The amount of heat that flows through one square foot of material in one hour at a 1°F temperature difference between its surfaces. Conductance values are given for a specific thickness of material. For homogeneous materials such as concrete, dividing its conductivity (k) by its thickness (X) gives the conductance (C).

Conduction: The process by which heat energy is transferred through materials (solids, liquids, or gases) by molecular excitation of adjacent molecules.

Conductivity (k): The amount of heat that flows through one square foot of material, one inch thick, in one hour, at a temperature difference of 1°F between its surfaces.

Convection: The transfer of heat within or by a moving fluid medium (liquid or gas).

Convective loop: A system for the transfer of heat from one point to another by convection. After losing or transferring the heat, the transfer medium returns to the source of heat to complete the cycle or loop.

Cooling pond: A body of water that dissipates heat by evaporation, convection, and radiation. Usually a fluid such as water.

Coolth tubes: Conduits used to cool a transfer fluid such as water or air by dissipating heat contained in the fluid to a heat sink surrounding the tubes.

Cross-ventilation: The flow of air through a building by virtue of openings located on opposite walls.

Dead air space: An air space that is sealed to prevent convective heat transfer into or out of the space.

Degree day (DD), cooling: see Degree day (DD) heating; the base temperature is generally established at 65°F by the national weather service. Cooling degree days are often measured above that base. For the thermal performance simulations in this book, 65°F was used.

Degree day (DD), heating: An expression of a climatic heating requirement expressed by the daily difference in degrees F below the average outdoor temperature for each day and an established indoor temperature base of 65°F. The total number of degree days over the heating season indicates the relative severity of the winter.

Demand limiter: A device that selectively switches off electrical equipment whenever total electrical demand rises beyond a predetermined level.

Density: The mass per unit volume of a substance, expressed in pounds per cubic foot.

Desiccant: A material which has the property of absorbing moisture from the air.

Differential thermostat: A control thermostat with two temperature sensors, typically one at the heat source and one at the storage site, to automatically control all or a part of a system.

Diffuse radiation: Radiation that has traveled an indirect path from the sun because it has been scattered by the particles in the atmosphere, such as air molecules, dust, and water vapor.

Direct (beam) radiation: Light that has traveled a straight path from the sun, as opposed to diffuse sky radiation.

Direct-gain system: Solar energy collected as heat and stored directly within a building.

Drum wall: A type of water wall using stacked drums for solar heat collection and storage.

Dry bulb temperature: A measure of air temperature when it is independent of radiation effects from surroundings, and when air motion relative to the measuring devices is not significant.

Earth tempering: The heating or cooling of a fluid or space by association with the ambient temperature of the adjacent earth.

Energy: The capacity for doing work; it may take a number of forms, all of which can be transformed from one into another: thermal (heat), mechanical (work), electrical, and chemical; in customary units, measured in kilowatt hours (kWh) or British thermal units (Btu).

Equinox: Either of the two times during a year when the sun crosses the celestial equator and when the length of the day and night are approximately equal—the autumnal equinox, on or about September 22, and the vernal equinox, on or about March 22.

Eutectic salts: Salts used for storing heat. At a given temperature, such salts melt, absorbing large amounts of heat; this will be released when the salts freeze. See Phase-change thermal storage.

Fahrenheit degrees (°F): A temperature scale in which the freezing point of water is set at 32°F and its boiling point at 212°F. Fahrenheit temperatures can be derived from centigrade temperatures by the equation $°F = 1.8 \times °C + 32$.

Fenestration: A term used to signify an opening in a building to admit light and/or air; windows.

Glazing: A covering of transparent or translucent material (glass or plastic) used for admitting light. Glazing retards heat losses from reradiation and convection. Examples: windows, skylights, greenhouse, collector coverings.

Glazing, double: A sandwich of two separated layers of glass or plastic enclosing air to create an insulating barrier. Sometimes the space is evacuated or filled with an inert gas to increase insulation value.

Greenhouse: See Sunspace.

Greenhouse effect: The characteristic tendency of some transparent materials, such as glass, to transmit shorter wavelength solar radiation (light) and absorb thermal radiation of longer wavelengths (heat), thus reducing heat loss and increasing heat gain.

Heat capacity: The amount of heat that a cubic foot of material can store with a one-degree increase in its temperature.

Heat gain: An increase in the amount of heat contained in a space, resulting from absorbed solar radiation or internal gain.

Heat loss: A decrease in the amount of heat contained in a space, resulting from heat flow through walls, windows, roof, and other building envelope components.

Heat-transfer medium: A medium, either liquid or gas, used to transport thermal energy.

Heat sink: A substance capable of accepting and storing heat.

Hybrid solar energy systems: Hybrid systems use passive design concepts in combination with active components (fan, pumps, etc.) for collection, storage, or distribution of energy.

Incident angle: The angle between incoming direct solar radiation falling on a surface and a line perpendicular (normal) to that surface.

Indirect-gain system: A solar heating system in which sunlight first strikes a thermal mass located between the sun and a living space. The sunlight absorbed by the mass is converted to heat and then transferred to the space.

Infiltration: The uncontrolled movement of outdoor air into the interior of a building through cracks around windows and doors or in walls, roofs, and floors.

Insolation: The total amount of solar radiation—direct, diffuse, and reflected—striking a surface exposed to the sky.

Insulating shade: Window or door shades with insulative properties to prevent heat loss or gain when closed.

Insulation: Materials or systems used to prevent heat loss or gain, usually employing very small dead-air spaces to limit conduction and/or convection, or reflective surfaces to minimize radiation.

Internal heat gain: An increase in the amount of heat contained within a space resulting from the energy given off by people, lights, equipment, machinery, pets, or other elements.

Isolated-gain system: A system in which solar collection and heat storage are isolated from the living spaces.

Latent heat: A change in heat content of a material that occurs without a corresponding change in temperature, accompanied by a change of state (for example, ice changing to water or water to steam). See Phase-change thermal storage.

Latitude: The angular distance north or south of the equator, measured in degrees of arc.

Mass wall: A wall composed of material with a relatively high thermal storage capability and conductance, such as concrete, masonry, or adobe. Usually placed inside of exterior glazing.

Mean radiant temperature: The weighted average surface temperature of walls, floors, ceilings.

MMBtu/yr.: Million British thermal units per year.

Night cooling: The cooling of a building or heat storage device by the radiation of excess heat into the night sky.

Night ventilation: The cooling of a building by night ventilation. Night ventilation can be achieved by forced means, such as a fan, or by natural means, such as the chimney effect or turbine vents.

Night ventilation cooling fraction: The percentage of the cooling load supplied by night ventilation in a solar house.

Opaque: Impenetrable by light.

Parasitic energy: The amount of energy derived from a depletable fuel source (coal, gas, oil, etc.) to run a solar system. For example, fans and pumps require parasitic energy.

Passive solar energy systems: Systems that rely upon the building's design and construction for collection, storage, and distribution of the sun's energy for heating and cooling. In a strict sense, such a system would use no energy other than the natural sources of sun and wind.

Percent of possible sunshine: The amount of time between sunrise and sunset that the sun is shining (not obscured by clouds), usually represented as a monthly or annual average.

Phase-change thermal storage: Materials which release or take in a quantity of heat when changing from solid to liquid or liquid to solid.

Photovoltaic cells: Devices for converting solar energy to electricity.

Radiation: The direct transfer of energy through space by means of electromagnetic waves.

Radiation, infrared: Electromagnetic radiation, whether from the sun or a warm body, with wavelengths longer than the red end of the visible spectrum (greater than 0.75 microns). We experience infrared radiation as heat; 53 percent of the radiation emitted by the sun is in the infrared band.

Radiation, solar: Electromagnetic radiation emitted by the sun.

Reflectance: The ratio of the amount of light reflected by a surface to the amount incident. What is not reflected is either absorbed by the material or transmitted through it. Good light reflectors are not necessarily good heat reflectors.

Relative humidity: The water content of moist air with respect to saturated air, expressed as a percentage.

Resistance (R): The opposite of conductivity.

Retrofit: Installation of solar water heating and/or solar heating or cooling systems in existing buildings not originally designed for that purpose.

R value: A unit of thermal resistance used for comparing insulating values of different materials; the higher the R value of a material, the greater its insulating properties.

Rock bed: See Rock bin.

Rock bin: Rocks placed in bins to store heat for later use.

Roof monitors: Skylights or glazing on the roof to admit solar energy and light to the interior of a building.

Roof-pond system: An indirect-gain heating and cooling system where the mass—water in plastic bags—is located on the roof of the space to be heated or cooled. As solar radiation heats the water, the heat is transferred into the space, usually through a metal-type support ceiling.

Selective surface: A surface or coating that has a high absorptance of incoming solar radiation but low emittance of longer wavelengths (heat).

Skylight: A clear or translucent panel set into a roof to admit sunlight into a building. Usually fixed, sometimes operable for venting.

Solar altitude: The angle of the sun above the horizon, measured in a vertical plane.

Solar aperture: An opening designed or placed primarily to admit solar energy into a space.

Solar chimney (also called a solar inducer): A type of solar collector with a dark absorption surface used to heat and move air. Usually located to receive summer sunlight. The heated air moves up and out the top of the chimney, causing an induction of outside or pretempered air into the building.

Solar collector: see Collector, solar.

Solar constant: The quantity of radiant solar heat received at the outer layer of the earth's atmosphere. Equal to 429 Btu/ft^2 hr ($\pm$ 1.5%).

Solar heating fraction: The percentage of the heating load supplied by solar energy.

Specific heat (Cp): The amount of heat required to raise the temperature of one pound of a substance 1°F in temperature. Units are Btu/lb.°F.

Stack effect: see Chimney effect. Usually associated with hot air's escaping from a building, increasing ventilation potential.

Stratification: In solar heating context, the formation of layers in a substance (fluid or gas) where the top layer is warmer than the bottom. Also called thermal stratification.

Sunspace: A glazed space attached to a building in which solar heat and light are collected for growing plants and/or space heating.

Thermal admittance (q): The amount of heat a square foot of surface will admit in one hour.

Thermal break (thermal barrier): An element of low heat conductivity placed to reduce or prevent heat flow.

Thermal flywheel (also called thermal inertia): The tendency of a building with large quantities of heavy materials to remain at the same temperature or to fluctuate only very slowly.

Thermal mass: The potential heat-storage capacity available in a given building. Mass walls, adobe, stone, brick, concrete, and water are examples of thermal mass.

Thermosiphon: The hydraulic system in which fluid circulation is caused by temperature differences. Warmer fluids expand and rise, cooler fluids contract and fall.

Translucent: Having the quality of transmitting light but causing sufficient diffusion to eliminate perception of distinct images.

Transmittance: The ratio of the radiant energy transmitted through a substance to the total radiant energy incident on its surface.

Transparent: Having the quality of transmitting light so that objects or images can be seen as if there were no intervening material.

Trombe wall: A masonry exterior wall (south facing) that collects and releases stored solar energy into a building by both radiant and convective means. This wall is insulated from the exterior by glass or other transparent material. Developed by Dr. Felix Trombe.

U-Value (overall coefficient of heat transfer): The amount of heat in Btu that flows through one square foot of roof, wall, or floor, in one hour, when there is a 1°F difference in temperature between the inside and outside air, under steady-state conditions. The U-value is the reciprocal of the total R-value. Units are Btu/hr.ft.^{2} °F.

Vapor barrier: A building material which inhibits the flow of moisture and air as well as minimizing condensation in walls, floors, and roofs.

Water wall: An interior wall of water-filled containers constituting a thermal storage mass.

Weather: The state of the atmosphere at a given time and place, described by such variables as temperature, moisture, wind velocity, and barometric pressure.

Weatherizing: The process of improving the properties of the weatherskin: increasing insulation, double-glazing, caulking, weather stripping, etc.

Weatherskin: The exterior of a building, separating the conditioned interior from the exterior.

Weather stripping: Narrow or jamb-width sections of thin metal or other material to prevent infiltration of air and moisture around windows and doors.

Wet bulb temperature: The lowest temperature attainable by evaporating water into the air without altering its energy content.

BIBLIOGRAPHY

A Survey of Passive Solar Homes. Washington, D.C.: AIA Research Corporation, 1980.

Anderson, Bruce and Michael Riordon. *The Solar Home Book.* Andover, Massachusetts: Brick House Publishing Company, 1976.

Anderson, Bruce and M. Wells. *Passive Solar Energy: The Homeowner's Guide to Natural Heating and Cooling.* Andover, Massachusetts: Brick House Publishing Company, 1981.

Antolini, Holly L., ed. *Sunset Homeowner's Guide to Solar Heating.* Menlo Park, California: Lane Publishing Company, 1978.

Aronin, Jeffrey Ellis. *Climate and Architecture.* New York: Reinhold Publishing Corporation, 1953.

ASHRAE Handbook of Fundamentals. New York: American Society of Heating, Refrigerating, and Air Conditioning Engineers, 1981.

Balcomb, Douglas, et al. *Passive Solar Design Handbook. Volume Two: Passive Solar Design Analysis.* Springfield, Virginia: US DOE/NTIS (# DOE/CS-0127/2), 1980.

Butti, Ken and J. Perlin. *A Golden Thread: 2500 Years of Solar Architecture and Technology.* New York: Van Nostrand Reinhold Company, 1980.

Caudill, William, et al. *A Bucket of Oil.* Boston: Cahners Books, 1974.

Clegg, P. and D. Watkins. *The Complete Greenhouse Book.* Charlotte, Vermont: Garden Way Publishing, 1978.

Climatic Atlas of the United States. Washington, D.C.: U.S. Department of Commerce, 1968.

Conklin, Groff and S. Blackwell Duncan. *The Weather-Conditioned House.* New York: Van Nostrand Reinhold, 1982.

Crowther, Richard and Solar Group Architects. *Sun/Earth.* Denver, Colorado: A. D. Hirshfeld Press, Inc., 1976. Updated and revised edition by Van Nostrand Reinhold, 1983.

Davis, A. J. and R. P. Schubert. *Alternative Natural Energy Sources in Building Design.* New York: Van Nostrand Reinhold Company, 1977.

Eccli, Eugene. *Low Cost Energy Efficient Shelter for the Owner and Builder.* Emmaus, Pennsylvania: Rodale Press, 1975.

Egan, David. *Concepts in Thermal Comfort.* Englewood Cliffs, New Jersey: Prentice-Hall, Inc., 1975.

Fisher, Rick and W. Yanda. *The Food and Heat Providing Solar Greenhouse: Design, Construction, Operation.* Santa Fe, New Mexico: John Muir Publications, 1976.

Giovoni, B. *Man, Climate and Architecture.* New York: Van Nostrand Reinhold Company, 1969.

Gropp, Louis. *Solar Houses: 48 Energy Saving Designs.* New York: Pantheon Books, 1978.

Johnson, Tim. *Solar Architecture: The Direct Gain Approach.* New York: McGraw-Hill, 1981.

Lebens, Ralph. *Passive Solar Heating Design.* New York: John Wiley and Sons, 1979.

Libbey-Owens-Ford. *Sun Angle Calculator.* Toledo: Libbey-Owens-Ford Company, 1974.

Mazria, Edward. *The Passive Solar Energy Book.* Emmaus, Pennsylvania: Rodale Press, 1979.

McClenon, Charles, ed. *Landscape Planning for Energy Conservation.* Reston, Virginia: Environmental Design Press, 1977.

McCullagh, J. C. *The Solar Greenhouse Book.* Emmaus, Pennsylvania: Rodale Press, 1978.

McGuinness, William, Benjamin Stein, and John Reynolds. *Mechanical and Electrical Equipment for Buildings, Sixth Edition.* New York: John Wiley and Sons, 1980.

Niles, Philip and K. Haggard. *Passive Solar Handbook for California.* Sacramento, California: California Energy Commission, 1980.

Oddo, S. *Solar Age Catalog.* Harrisville, New Hampshire: Solar Vision, Inc., 1977.

Olgyay, Aladan and V. Olgyay. *Solar Control and Shading Devices.* Princeton, New Jersey: Princeton University Press, 1967.

Olgyay, Victor. *Design with Climate.* Princeton, New Jersey: Princeton University Press, 1963.

Packard, Robert T., ed. *Architectural Graphic Standards.* Seventh Edition. New York: John Wiley and Sons, 1981.

Rapoport, A. *House Form and Culture.* Englewood Cliffs, New Jersey: Prentice-Hall, Inc., 1969.

Regional Guidelines for Building Passive Energy Conserving Homes. Washington, D.C.: AIA Research Corporation, 1978.

Robinette, G. O. *Plants, People and Environmental Quality.* Washington, D.C.: U.S. Department of the Interior, National Park Service, 1972.

Shurcliff, William A. *Thermal Shutters and Shades.* Andover, Massachusetts: Brick House Publishing Co., 1980.

———. *Solar Heated Buildings of North America: 120 Outstanding Examples.* Andover, Massachusetts: Brick House Publishing Company, 1978.

Underground Space Center, University of Minnesota. *Earth Sheltered Housing Design.* New York: Van Nostrand Reinhold, 1979.

———. *Earth Sheltered Community Design: Energy-Efficient Residential Development.* New York: Van Nostrand Reinhold, 1981.

———. *Earth Sheltered Homes: Plans and Designs.* New York: Van Nostrand Reinhold, 1981.

Wagner, Walter F., Jr., ed. *Energy-Efficient Buildings.* New York: McGraw-Hill Book Company, 1980.

Watson, Donald, ed. *Energy Conservation Through Building Design.* New York: McGraw-Hill Book Company, 1979.

Watson, Donald and K. Labs. *Climatic Design for Home Building.* Guilford, Connecticut: Donald Watson, 1980.

Wright, David. *Natural Solar Architecture: A Passive Primer.* New York: Van Nostrand Reinhold Company, 1978.

INDEX